Lecture Notes in Mathematics

Edited by A. Dold, Heidelberg and B. Eckmann, Zürich

Series: University of Maryland, College Park
Advisor: L. Greenberg

388

Ronald L. Lipsman

Group Representations

Springer-Verlag
Berlin · Heidelberg · New York

Lecture Notes in Mathematics

Vols. 1-149 are also available. For further information, please contact your book-seller or Springer-Verlag.

Vol. 150: Heyer, Dualität lokalkompakter Gruppen. XIII, 372 Seiten. 1970. DM 20,-

Vol. 151: M. Demazure et A. Grothendieck, Schémas en Groupes I. (SGA 3). XV, 562 pages. 1970. DM 24,-

Vol. 152: M. Demazure et A. Grothendieck, Schémas en Groupes II. (SGA 3). IX, 654 pages. 1970. DM 24,-

Vol. 153: M. Demazure et A. Grothendieck, Schémas en Groupes III. (SGA 3). VIII, 529 pages. 1970. DM 24,-

Vol. 154: A. Lascoux et M. Berger, Variétés Kähleriennes Compactes. VII, 83 pages. 1970. DM 16,-

Vol. 155: Several Complex Variables I, Maryland 1970. Edited by J. Horváth. IV, 214 pages. 1970. DM 18,-

Vol. 156: R. Hartshorne, Ample Subvarieties of Algebraic Varieties. XIV, 256 pages. 1970. DM 20,-

Vol. 157: T. tom Dieck, K. H. Kamps und D. Puppe, Homotopietheorie. VI, 256 Seiten. 1970. DM 20,-

Vol. 158: T. G. Ostrom, Finite Translation Planes. IV. 112 pages. 1970. DM 16,-

Vol. 159: R. Ansorge und R. Hass. Konvergenz von Differenzenverfahren für lineare und nichtlineare Anfangswertaufgaben. VIII, 145 Seiten. 1970. DM 16,-

Vol. 160: L. Sucheston, Contributions to Ergodic Theory and Probability. VII, 277 pages. 1970. DM 20,-

Vol. 161: J. Stasheff, H-Spaces from a Homotopy Point of View. VI, 95 Pages. 1970. DM 16,-

Vol. 162: Harish-Chandra and van Dijk, Harmonic Analysis on Reductive p-adic-Groups. IV, 125 pages. 1970. DM 16,-

Vol. 163: P. Deligne, Equations Differentielles à Points Singuliers Reguliers. III, 133 pages. 1970. DM 16,-

Vol. 164: J. P. Ferrier, Séminaire sur les Algèbres Complètes. II, 69 pages. 1970. DM 16,-

Vol. 165: J. M. Cohen, Stable Homotopy. V, 194 pages. 1970. DM 16,-

Vol. 166: A. J. Silberger, PGL_2 over the p-adics: its Representations, Spherical Functions, and Fourier Analysis. VII, 202 pages. 1970. DM 18,-

Vol. 167: Lavrentiev, Romanov and Vasiliev, Multidimensional Inverse Problems for Differential Equations. V, 59 pages. 1970. DM 16,-

Vol: 168: F. P. Peterson, The Steenrod Algebra and its Applications: A Conference to Celebrate N. E. Steenrod's Sixtieth Birthday. VII, 317 pages. 1970. DM 22,-

Vol. 169: M. Raynaud, Anneaux Locaux Henséliens. V, 129 pages. 1970. DM 16,-

Vol. 170: Lectures in Modern Analysis and Applications III. Edited by C. T. Taam. VI, 213 pages. 1970. DM 18,-

Vol. 171: Set-Valued Mappings, Selections and Topological Properties of 2^X. Edited by W. M. Fleischman. X, 110 pages. 1970. DM 16,-

Vol. 172: Y.-T. Siu and G. Trautmann, Gap-Sheaves and Extension of Coherent Analytic Subsheaves. V, 172 pages. 1971. DM 16,-

Vol. 173: J. N. Mordeson and B. Vinograde, Structure of Arbitrary Purely Inseparable Extension Fields. IV, 138 pages. 1970. DM 16,-

Vol. 174: B. Iversen, Linear Detterminants with Applications to the Picard Scheme of a Family of Algebraic Curves. VI, 69 pages. 1970. DM 16,-

Vol. 175: M. Brelot, On Topologies and Boundaries in Potential Theory. VI, 176 pages. 1971. DM 18,-

Vol. 176: H. Popp, Fundamentalgruppen algebraischer Mannigfaltigkeiten. IV, 154 Seiten. 1970. DM 16,-

Vol.177: J. Lambek, Torsion Theories, Additive Semantics and Rings of Quotients. VI, 94 pages. 1971. DM 16,-

Vol. 178: Th. Bröcker und T. tom Dieck, Kobordismentheorie. XVI, 191 Seiten. 1970. DM 18,-

Vol. 179: Séminaire Bourbaki - vol. 1968/69. Exposés 347-363. IV. 295 pages. 1971. DM 22,-

Vol. 180: Séminaire Bourbaki - vol. 1969/70. Exposés 364-381. IV, 310 pages. 1971. DM 22,-

Vol. 181: F. DeMeyer and E. Ingraham, Separable Algebras over Commutative Rings. V, 157 pages. 1971. DM 16,-

Vol. 182: L. D. Baumert. Cyclic Difference Sets. VI, 166 pages. 1971. DM 16,-

Vol. 183: Analytic Theory of Differential Equations. Edited by P. F. Hsieh and A. W. J. Stoddart. VI, 225 pages. 1971. DM 20,-

Vol. 184: Symposium on Several Complex Variables, Park City, Utah, 1970. Edited by R. M. Brooks. V, 234 pages. 1971. DM 20,-

Vol. 185: Several Complex Variables II, Maryland 1970. Edited by J. Horváth. III, 287 pages. 1971. DM 24,-

Vol. 186: Recent Trends in Graph Theory. Edited by M. Capobianco/ J. B. Frechen/M. Krolik. VI, 219 pages. 1971. DM 18,-

Vol. 187: H. S. Shapiro, Topics in Approximation Theory. VIII, 275 pages. 1971. DM 22,-

Vol. 188: Symposium on Semantics of Algorithmic Languages. Edited by E. Engeler. VI, 372 pages. 1971. DM 26,-

Vol. 189: A. Weil, Dirichlet Series and Automorphic Forms. V, 164 pages. 1971. DM 16,-

Vol. 190: Martingales. A Report on a Meeting at Oberwolfach, May 17-23, 1970. Edited by H. Dinges. V, 75 pages. 1971. DM 16,-

Vol. 191: Séminaire de Probabilités V. Edited by P. A. Meyer. IV, 372 pages. 1971. DM 26,-

Vol. 192: Proceedings of Liverpool Singularities - Symposium I. Edited by C. T. C. Wall. V, 319 pages. 1971. DM 24,-

Vol. 193: Symposium on the Theory of Numerical Analysis. Edited by J. Ll. Morris. VI, 152 pages. 1971. DM 16,-

Vol. 194: M. Berger, P. Gauduchon et E. Mazet. Le Spectre d'une Variété Riemannienne. VII, 251 pages. 1971. DM 22,-

Vol. 195: Reports of the Midwest Category Seminar V. Edited by J. W. Gray and S. Mac Lane. III, 255 pages. 1971. DM 22,-

Vol. 196: H-spaces - Neuchâtel (Suisse)- Août 1970. Edited by F. Sigrist. V, 156 pages. 1971. DM 16,-

Vol. 197: Manifolds - Amsterdam 1970. Edited by N. H. Kuiper. V, 231 pages. 1971. DM 20,-

Vol. 198: M. Hervé, Analytic and Plurisubharmonic Functions in Finite and Infinite Dimensional Spaces. VI, 90 pages. 1971. DM 16,-

Vol. 199: Ch. J. Mozzochi, On the Pointwise Convergence of Fourier Series. VII, 87 pages. 1971. DM 16,-

Vol. 200: U. Neri, Singular Integrals. VII, 272 pages. 1971. DM 22,-

Vol. 201: J. H. van Lint, Coding Theory. VII, 136 pages. 1971. DM 16,-

Vol. 202: J. Benedetto, Harmonic Analysis in Totally Disconnected Sets. VIII, 261 pages. 1971. DM 22,-

Vol. 203: D. Knutson, Algebraic Spaces. VI, 261 pages. 1971. DM 22,-

Vol. 204: A. Zygmund, Intégrales Singulières. IV, 53 pages. 1971. DM 16,-

Vol. 205: Séminaire Pierre Lelong (Analyse) Année 1970. VI, 243 pages. 1971. DM 20,-

Vol. 206: Symposium on Differential Equations and Dynamical Systems. Edited by D. Chillingworth. XI, 173 pages. 1971. DM 16,-

Vol. 207: L. Bernstein, The Jacobi-Perron Algorithm - Its Theory and Application. IV, 161 pages. 1971. DM 16,-

Vol. 208: A. Grothendieck and J. P. Murre, The Tame Fundamental Group of a Formal Neighbourhood of a Divisor with Normal Crossings on a Scheme. VIII, 133 pages. 1971. DM 16,-

Vol. 209: Proceedings of Liverpool Singularities Symposium II. Edited by C. T. C. Wall. V, 280 pages. 1971. DM 22,-

Vol: 210: M. Eichler, Projective Varieties and Modular Forms. III, 118 pages. 1971. DM 16,-

Vol. 211: Théorie des Matroïdes. Edité par C. P. Bruter. III, 108 pages. 1971. DM 16,-

Vol. 212: B. Scarpellini, Proof Theory and Intuitionistic Systems. VII, 291 pages. 1971. DM 24,-

Vol. 213: H. Hogbe-Nlend, Théorie des Bornologies et Applications. V, 168 pages. 1971. DM 18,-

Vol. 214: M. Smorodinsky, Ergodic Theory, Entropy. V, 64 pages. 1971. DM 16,-

continuation on page 173

Lecture Notes in Mathematics

Edited by A. Dold, Heidelberg and B. Eckmann, Zürich
Series: University of Maryland, College Park
Advisor: L. Greenberg

388

Ronald L. Lipsman
University of Maryland, College Park, MD/USA

Group Representations

A Survey of Some Current Topics

Springer-Verlag
Berlin · Heidelberg · New York 1974

AMS Subject Classifications (1970): 22-02, 43-02

ISBN 3-540-06790-6 Springer-Verlag Berlin · Heidelberg · New York
ISBN 0-387-06790-6 Springer-Verlag New York · Heidelberg · Berlin

Offsetdruck: Julius Beltz, Hemsbach/Bergstr.

To

Irving, Blanche, Lenny and Barry

PREFACE

The idea for these notes was conceived as I watched the progress of several thesis students in my department at the University of Maryland. Although they were all working in representation theory, and although they talked together regularly, each seemed to have a limited idea of what the other was doing. In particular, while one was into semisimple Lie groups, another was involved with nilpotent Lie groups and there was a minimum of mutual understanding. Now, it is true that even the most advanced researchers in these areas have undertaken too little intercommunication; so, it is not at all surprising that students in the field should fare no better.

With an eye towards remedying this unpleasant situation (at least locally), I conducted a survey course in group representations during the spring of 1973. The main goal was to open the students eyes to various different vistas within the general panorama of group representations. A secondary goal, motivated in part by my current work, was to present the various interactions and connections between these fields (as it has rarely if ever been done). A third and final goal was to help the students in their own work by providing numerous examples and exercises, indications of current problems in the field, and copious bibliographical references.

I think the course had a fair degree of success. In any event, several students and faculty have encouraged me to put it in writing to see if anyone else might be interested. The result is these notes. It is my hope that they might prove useful to students trying to learn the field, workers in one area of representation theory who lack familiarity with another, or as a general source of reference.

The topics treated are the following. In Chapter I, I present an introduction to the representation theory of semisimple Lie groups. This is a beautiful and elaborate subject -- due in large part to Harish-Chandra -- and to tell the full story would fill several books (if you don't believe me, ask Garth Warner). I have tried to hit the highlights in a consistent and lively manner. It is my feeling that the current attractiveness of group representations to the mathematical community is due in part to the compelling beauty and power of this particular theory. It is also due to the excellent foundations laid for the theory by G. Mackey a generation ago. That is the subject matter of Chapter II. In order to appreciate Chapter II (and the rest of the book for that matter), the reader needs a good knowledge of induced representations. (In Appendix A, the reader will find a short introduction to induced representations which includes the important basic definitions and properties.) In the many examples that are found in Chapter II (especially in section B), I have tried to present what I feel is one of the few systematic attempts to relate many of Mackey's original results on induced representations to the particular case of semisimple groups.

Chapter III is devoted to Mackey's theory of representations of group extensions. This lovely theory is a natural outgrowth of the material in Chapter II and I have tried to present it that way. Once again, I offer many examples, some of which serve as motivation for the subject matter of Chapter V. I also include in Chapter III the theory (recently developed by Adam Kleppner and myself) of Plancherel measure for group extensions. The Imprimitivity Theorem (discussed in Appendix B) is an indispensable tool in this chapter.

In Chapter IV, I give the story of another splendid success in group representations - the theory of orbits and the representations of simply connected nilpotent Lie groups. The main results are due to Dixmier, Pukanszky and especially Kirillov. In Chapter V, under the general heading of algebraic groups, I tried to do two things. First, I indicated how previously described results on Lie groups go over to p-adic groups (sometimes they don't, unfortunately). The main results are due to Harish-Chandra, Moore and others. Secondly, I tried to suggest to what extent the Mackey theory of group extensions might be applied to interconnect the (heretofore divergent) areas of semisimple and unipotent groups. This work is yet very young, and I feel will receive much attention in the future.

Finally in Chapter VI I give a very brief introduction to the representation theory of solvable groups. For real groups the results are due to Auslander and Kostant, for p-adic groups to Howe.

Although I have covered a broad spectrum of results in representation theory here, I have by no means covered it all. Perhaps the single most important topic omitted is the multiplicity theory of the regular representation of a group G acting on $L_2(G/\Gamma)$, Γ a discrete subgroup. There are other omissions of course as well. Most were dictated by considerations of time. It was only a one-semester course, and I didn't want to get old writing up these notes. So much for errors of omission. As for the other kind, I apologize flat out. If there are any, they are due to laziness, ignorance or my own peculiar way of looking at things.

These notes run very close to the course as originally presented in my lectures. Thus the reader will find many theorems stated without proof, some stated with partial proofs or indications of proof, and a few actually proved in entirety (nobody's perfect). I'm sorry to say that I had no grand scheme for deciding how much proof to supply for any particular theorem -- to a great extent it depended on the day-to-day needs of time as the course progressed. Another word of caution. Sometimes group actions appear on the left, at other times on the right. Don't search for hidden significance -- again convenience dictated the convention.

A word on prerequisites for reading these notes. The reader is expected to have a non-trivial knowledge of representation theory. For example, he should know things that are in the books by Naimark, Loomis, Dixmier (second half of C^*-algebras, anyway), and the introduction to the Auslander-Moore Memoir. In addition, he is assumed to be familiar with Lie groups and Lie algebras, Borel spaces, operator theory, algebraic varieties (a little bit), and a few other things on occasion. Terminology and notation that is not defined in the main text can usually be found in the listing at the back of these notes.

Finally, it is my pleasure to thank Eloise Carlton, Robert Martin, John Pesek, and William Rapley. They were a hardy bunch to have put up with me for fourteen weeks. I am also grateful for the help and encouragement given to me by my colleagues Leon Greenberg and Adam Kleppner, and for the excellent typing job done by Debbie Curran and Betty Vanderslice.

College Park, Maryland

Fall 1973

CONTENTS

CHAPTER I. INTRODUCTION TO SEMISIMPLE THEORY

A. BASIC STRUCTURE

The following material is a summary (taken mostly from Helgason [1]) of subject matter one needs to know in order to get started in the representation theory of semisimple Lie groups.

Let k be a field of characteristic zero, $\mathfrak{g}$ a Lie algebra over k. The *adjoint representation* of $\mathfrak{g}$ is given by

$$\text{ad}: \mathfrak{g} \to \text{End}\,\mathfrak{g} \qquad \text{ad}\,X(Y) = [X,Y], \quad X,Y \in \mathfrak{g}.$$

Set $B(X,Y) = \text{tr}(\text{ad}\,X\,\text{ad}\,Y)$, the so-called *Cartan-Killing form.* $\mathfrak{g}$ is called *semisimple* if B is non-degenerate. $\mathfrak{g}$ is *simple* if it is semisimple and has no non-trivial ideals. It's easy to check that $B(X,\text{ad}\,Z(Y)) = -B(\text{ad}\,Z(X),Y)$, so that ad is skew symmetric with respect to B.

Now assume $k = \mathbb{C}$ and $\mathfrak{g}$ is a semisimple Lie algebra over $\mathbb{C}$. By a *Cartan subalgebra* of $\mathfrak{g}$ we mean a subalgebra $\mathfrak{h} \subseteq \mathfrak{g}$ which is maximal abelian and has the property that $\text{ad}\,H$ $(H \in \mathfrak{h})$ is a semisimple endomorphism of $\mathfrak{g}$ (that is diagonalizable over $\mathbb{C}$).

EXERCISE. Show that $\text{ad}(\mathfrak{h})$ is a maximal set of simultaneously diagonalizable elements of $\text{ad}(\mathfrak{g})$.

Next if V is a vector space over $\mathbb{C}$, $A \in \text{End}(V)$, and $\alpha_1,\ldots,\alpha_r$ are the eigenvalues of A, we set $V_i = \{v \in V: (A - \alpha_i I)^m v = 0$, some $m \geqq 1\}$. Then: (1) $V = \sum_{i=1}^{r}{}^{\oplus}\, V_i$ and each V_i is A-invariant; (2) $A = S + N$ where S is semisimple, N is nilpotent, $SN = NS$ and $S(\sum_{i=1}^{r} v_i) = \sum_{i=1}^{r} \alpha_i v_i$, $v_i \in V_i$; (3) $\det(\lambda I - A) = \prod_{i=1}^{r} (\lambda - \alpha_i)^{d_i}$,

$d_i = \dim V_i$.

We apply the preceding to ad H, $H \in \mathfrak{g}$, as an endomorphism of $\mathfrak{g}$. Let $0 = \lambda_0, \lambda_1, \ldots, \lambda_r$ denote the eigenvalues. For $\lambda \in \mathbb{C}$, set $\mathfrak{g}(H,\lambda) = \{X \in \mathfrak{g} : (\text{ad } H - \lambda I)^m X = 0, \text{ some } m \geqq 1\}$. Then $\mathfrak{g} = \sum_{i=0}^{r} {}^{\oplus} \mathfrak{g}(H,\lambda_i)$. Call H *regular* if $\dim \mathfrak{g}(H,0)$ has a minimum value.

THEOREM 1. *Let* $H_0 \in \mathfrak{g}$ *be regular. Then* $\mathfrak{g}(H_0,0)$ *is a Cartan subalgebra of* $\mathfrak{g}$.

Now let $\mathfrak{g}$ be a complex semisimple Lie algebra, $\mathfrak{h} \subseteq \mathfrak{g}$ a Cartan subalgebra. For $\alpha \in \text{Hom}_{\mathbb{C}}(\mathfrak{h},\mathbb{C})$, set $\mathfrak{g}^{\alpha} = \{X \in \mathfrak{g} : \text{ad } H(X) = \alpha(H)X, \ \forall H \in \mathfrak{h}\}$. It follows from the Jacobi identity that $[\mathfrak{g}^{\alpha}, \mathfrak{g}^{\beta}] \subseteq \mathfrak{g}^{\alpha+\beta}$. α is called a *root* if $\alpha \neq 0$ and $\mathfrak{g}^{\alpha} \neq \{0\}$. Denote the set of roots by Δ.

THEOREM 2. (i) $\mathfrak{g} = \mathfrak{h} + \sum_{\alpha \in \Delta} \mathfrak{g}^{\alpha}$.

(ii) $\dim \mathfrak{g}^{\alpha} = 1$ *if* $\alpha \in \Delta$.

(iii) $\alpha, \beta \in \Delta$, $\alpha + \beta \neq 0 \Rightarrow B(\mathfrak{g}^{\alpha}, \mathfrak{g}^{\beta}) = 0$.

(iv) $B|_{\mathfrak{h} \times \mathfrak{h}}$ *is non-degenerate; thus if* $\alpha \in \Delta$, *there is a unique* $H_{\alpha} \in \mathfrak{h}$ *such that* $B(H,H_{\alpha}) = \alpha(H)$, $H \in \mathfrak{h}$.

(v) *If* $\alpha \in \Delta$, *then* $-\alpha \in \Delta$ *and* $[\mathfrak{g}^{\alpha}, \mathfrak{g}^{-\alpha}] = \mathbb{C}H_{\alpha}$. *Indeed*

$$[X_{\alpha}, X_{-\alpha}] = B(X_{\alpha}, X_{-\alpha})H_{\alpha}, \quad X_{\alpha} \in \mathfrak{g}^{\alpha}, \quad X_{-\alpha} \in \mathfrak{g}^{-\alpha}.$$

EXERCISES. (1) Take $\mathfrak{g} = sl(n,\mathbb{C})$ = the $n \times n$ complex matrices of trace zero, $\mathfrak{h}$ = the subalgebra of diagonal matrices. Show that $\mathfrak{h}$ is a Cartan subalgebra, that $\Delta = \{H \to e_i(H) - e_j(H)$ if $H = \text{diag}(e_1(H), e_2(H), \ldots, e_n(H))\}$, and that $B(X,Y) = 2n \text{ tr}(XY)$.

(2) Take $\mathfrak{g} = sp(n,\mathbb{C}) = \{X \in M(2n,\mathbb{C}) : JX + {}^tXJ = 0, \ J = \begin{pmatrix} 0 & I \\ -I & 0 \end{pmatrix}\}$.

Then a Cartan subalgebra $\mathfrak{h}$ is spanned by the vectors $E_{i,i} - E_{n+i,n+i}$, $i \leqq i \leqq n$. Show that the Killing form is $B(X,Y) = (2n+2)\mathrm{tr}(XY)$.

THEOREM 3. *Let* $\alpha,\beta \in \Delta$. *Call the* α*-series containing* β *the set of all roots of the form* $\beta + n\alpha$, $n \in \mathbb{Z}$.

(i) *The* α*-series containing* β *has the form* $\{\beta + n\alpha\colon p \leqq n \leqq q\}$ *where* $\frac{-2\beta(H_\alpha)}{\alpha(H_\alpha)} = p + q$.

(ii) *If* $X_\alpha \in \mathfrak{g}^\alpha$, $X_\beta \in \mathfrak{g}^\beta$, $X_{-\alpha} \in \mathfrak{g}^{-\alpha}$, *then* $[X_{-\alpha},[X_\alpha,X_\beta]] = \frac{q(1-p)}{2}\alpha(H_\alpha)B(X_\alpha,X_{-\alpha})X_\beta$.

(iii) *If* $r\alpha$ *is a root,* $r \in \mathbb{C}$, *then* $r = 0$, 1, *or* -1.

(iv) *Suppose* $\alpha + \beta \neq 0$. *Then* $[\mathfrak{g}^\alpha, \mathfrak{g}^\beta] = \mathfrak{g}^{\alpha+\beta}$.

(v) $B|_{\mathfrak{h}^*\times\mathfrak{h}^*}$ *is strictly positive definite, where* $\mathfrak{h}^* = \sum_{\alpha\in\Delta} \mathbb{R}H_\alpha$.

(vi) $\mathfrak{h} = \mathfrak{h}^* + i\mathfrak{h}^*$, *a direct sum.*

(vii) $X_\alpha \in \mathfrak{g}^\alpha$ *can be chosen so that* $[X_\alpha,X_{-\alpha}] = H_\alpha$, $\alpha \in \Delta$, $[X_\alpha,X_\beta] = 0$ *if* $\alpha + \beta \notin \Delta$, $\alpha + \beta \neq 0$, *and* $[X_\alpha,X_\beta] = N_{\alpha,\beta}X_{\alpha+\beta}$, $\alpha + \beta \in \Delta$, *where* $N_{\alpha,\beta} = -N_{-\alpha,-\beta}$ *is real. (In fact, according to a famous theorem of Chevalley* [1], *the* X_α *can be chosen so that the* $N_{\alpha,\beta}$ *are actually integral.)*

All complex simple algebras were classified by Cartan (see e.g. Helgason [1] or Humphreys [1]). The so-called classical cases are $sl(n,\mathbb{C})$, $sp(n,\mathbb{C})$ and $so(n,\mathbb{C})$. We have met the first two in the preceding exercise. The last is $so(n,\mathbb{C}) = \{X \in M(n,\mathbb{C})\colon {}^tX = -X\}$ (actually in the classification, it is common and appropriate to subdivide into even and odd n). As corresponding Lie groups we may take $SL(n,\mathbb{C}) = \{g \in GL(n,\mathbb{C})\colon \det g = 1\}$, $Sp(n,\mathbb{C}) = \{g \in GL(2n,\mathbb{C})\colon {}^tgJg = J$, $J = \begin{pmatrix} 0 & I \\ -I & 0 \end{pmatrix}\}$, and $SO(n,\mathbb{C}) = \{g \in G\colon g = {}^tg^{-1}\}$. It is these and suitable "real forms" of these groups that we shall meet most commonly in our study of semisimple Lie groups.

Now write $\mathfrak{g}^R$ for $\mathfrak{g}$ with scalars restricted to $\mathbb{R}$. $\mathfrak{g}^R$ is a real Lie algebra. Also denote the map $X \to iX$ on $\mathfrak{g}$ by J. Then a subalgebra $\mathfrak{g}_0 \subseteq \mathfrak{g}^R$ is called a *real form* of $\mathfrak{g}$ if $\mathfrak{g}^R = \mathfrak{g}_0 + J\mathfrak{g}_0$ is a direct sum.

EXERCISE. Check that (1) $sl(n,\mathbb{R})$ = the $n \times n$ real matrices of trace zero, and (2) $su(n)$ = the $n \times n$ complex skew hermitian matrices of trace zero are both real forms of $sl(n,\mathbb{C})$.

THEOREM 4. *Every complex semisimple Lie algebra has both a compact real form and a normal real form.*

The meaning of these terms is the following. $\mathfrak{g}_0$ is a *normal real form* of $\mathfrak{g}$ if there is $\mathfrak{h}_0 \subseteq \mathfrak{g}_0$ such that $\mathfrak{h}_0$ is a maximal abelian subalgebra of $\mathfrak{g}_0$, ad $\mathfrak{h}_0$ is diagonalizable over $\mathbb{R}$, and $\mathfrak{h} = \mathfrak{h}_0 + i\mathfrak{h}_0$ is a Cartan subalgebra of $\mathfrak{g}$. In fact, in the previous notation, $\mathfrak{g}_0 = \mathfrak{h}^* + \sum_{\alpha\in\Delta} \mathbb{R}X_\alpha$ is a normal real form of $\mathfrak{g}$. Next, a semisimple Lie algebra $\mathfrak{a}$ over $\mathbb{R}$ is called *compact* if its adjoint group $Ad(\mathfrak{a})$ (i.e., the analytic subgroup of $GL(\mathfrak{a})$ corresponding to the Lie algebra $ad(\mathfrak{a})$) is compact. A compact real form $\mathfrak{u}$ of $\mathfrak{g}$ can always be obtained by setting $\mathfrak{u} = i\mathfrak{h}^* + \sum_{\alpha\in\Delta^+} \{zX_\alpha - \bar{z}X_{-\alpha}: z \in \mathbb{C}\}$, where Δ^+ denotes the positive roots (for some ordering).

EXERCISE. Show that for $\mathfrak{g} = sl(n,\mathbb{C})$, we have $\mathfrak{g}_0 = sl(n,\mathbb{R})$ and $\mathfrak{u} = su(n)$.

We now change notation and write $\mathfrak{g}$ for a real semisimple Lie algebra. If needed, $\mathfrak{g}_c$ will denote its complexification.

DEFINITION. A Lie algebra $\mathfrak{g}$ over $\mathbb{R}$ is called *reductive* if ad $\mathfrak{g}$ is completely reducible.

LEMMA 5. *Let* $\mathfrak{g}$ *be reductive,* $\mathfrak{c}$ = Cent $\mathfrak{g}$. *Then* $\mathfrak{g} = \mathfrak{c} \oplus [\mathfrak{g},\mathfrak{g}]$ *and* $[\mathfrak{g},\mathfrak{g}]$ *is semisimple.*

Proof (Indication). Since ad is completely reducible, we can write $\mathfrak{g} = \oplus \mathfrak{g}_i$, where the $\mathfrak{g}_i$ are irreducible subspaces of $\mathfrak{g}$ under ad. In particular the $\mathfrak{g}_i$ are ideals in $\mathfrak{g}$ and $[\mathfrak{g}_i, \mathfrak{g}_j] = \{0\}$, $i \neq j$. Set $\mathfrak{a} = \oplus \mathfrak{g}_i$: dim $\mathfrak{g}_i = 1$, and $\mathfrak{b} = \oplus \mathfrak{g}_i$: dim $\mathfrak{g}_i > 1$. Clearly the latter $\mathfrak{g}_i$ are simple, and so $\mathfrak{b}$ is semisimple. $\mathfrak{a}$ is abelian and $\mathfrak{g} = \mathfrak{a} \oplus \mathfrak{b}$. It is easy to check that $\mathfrak{a} = \mathfrak{c}$ and $\mathfrak{b} = [\mathfrak{g},\mathfrak{g}]$.

EXAMPLE. If $\mathfrak{g} = gl(n,\mathbb{R}) = M(n,\mathbb{R})$, then $[\mathfrak{g},\mathfrak{g}] = sl(n,\mathbb{R})$ and $\mathfrak{c}$ is the central subalgebra of diagonal matrices with a common entry.

THEOREM 6. *Let* $\mathfrak{g}$ *be a real semisimple Lie algebra. Then*

(i) $\mathfrak{g}$ *is compact if and only if* B *is strictly negative-definite.*

(ii) *If* $\mathfrak{g}$ *is compact, then the universal covering group of* Ad($\mathfrak{g}$) *is compact.*

(iii) *Any compact connected solvable Lie group is a torus, and so any compact Lie group has a reductive Lie algebra.*

EXERCISES. (1) Prove (i) and (iii).

(2) Show that groups corresponding to the compact real forms of the classical groups are:

$$SL(n,\mathbb{C}) \longleftrightarrow SU(n) = \{g \in SL(n,\mathbb{C}): \bar{g} = {}^t g^{-1}\}$$

$$Sp(n,\mathbb{C}) \longleftrightarrow Sp(n) = \{g \in Sp(n,\mathbb{C}): \bar{g} = {}^t g^{-1}\}$$

$$SO(n,\mathbb{C}) \longleftrightarrow SO(n) = \{g \in SO(n,\mathbb{C}): g = \bar{g} = {}^t g^{-1}\}.$$

SU(n) and Sp(n) are simply connected, while SO(n), $n \geqq 3$, has a two-fold simply connected covering group (usually denoted Spin(n)).

By a *Cartan involution* we mean an automorphism θ of $\mathfrak{g}$ with the property that $\theta^2 = 1$ and $B_\theta(X,Y) = -B(X,\theta Y)$ is strictly positive-definite. By a *Cartan decomposition* we mean a direct sum decomposition $\mathfrak{g} = \mathfrak{k} + \mathfrak{p}$ such that the map $\theta\colon \mathfrak{g} \to \mathfrak{g}$ given by $X + Y \to X - Y$, $X \in \mathfrak{k}$, $Y \in \mathfrak{p}$, is a Cartan involution. Clearly then $\mathfrak{k} = \{X \in \mathfrak{g} : \theta X = X\}$, $\mathfrak{p} = \{Y \in \mathfrak{g} : \theta Y = -Y\}$.

EXAMPLE. $\mathfrak{g} = sl(n,\mathbb{R})$, $\theta X \to -{}^tX$. It's easily verified that this is a Cartan involution, $\mathfrak{k} = so(n) =$ the $n \times n$ real skew symmetric matrices of trace zero, $\mathfrak{p} =$ the $n \times n$ symmetric matrices of trace zero.

THEOREM 7. *Every semisimple Lie algebra* $\mathfrak{g}$ *over* $\mathbb{R}$ *has a Cartan decomposition and any two are conjugate.*

By conjugate we mean that if $\mathfrak{g} = \mathfrak{k}_1 + \mathfrak{p}_1 = \mathfrak{k}_2 + \mathfrak{p}_2$, then there is $x \in \mathrm{Ad}(\mathfrak{g})$ such that $\mathrm{Ad}\, x(\mathfrak{k}_1) = \mathfrak{k}_2$, $\mathrm{Ad}\, x(\mathfrak{p}_1) = \mathfrak{p}_2$.

It's worth noting here that $[\mathfrak{k},\mathfrak{k}] \subseteq \mathfrak{k}$, $[\mathfrak{k},\mathfrak{p}] \subseteq \mathfrak{p}$ and $[\mathfrak{p},\mathfrak{p}] \subseteq \mathfrak{k}$. (These follow easily from $\theta[X,Y] = [\theta X,\theta Y]$.) In particular, $\mathfrak{k}$ is a subalgebra, $\mathfrak{p}$ is a $\mathfrak{k}$-module and $\mathfrak{p}$ is not an algebra. It also follows easily that with respect to the form B_θ we have: $\mathrm{ad}\, X$ $(X \in \mathfrak{k})$ is skew symmetric; $\mathrm{ad}\, Y$ $(Y \in \mathfrak{p})$ is symmetric.

EXERCISE. Show that $\mathfrak{k}$ is a compactly embedded subalgebra of $\mathfrak{g}$. That means that the analytic subgroup of $\mathrm{Ad}(\mathfrak{g})$ corresponding to $\mathrm{Ad}(\mathfrak{k})$ is a compact group.

THEOREM 8. *Let* $\mathfrak{g}$ *be a semisimple Lie algebra over* $\mathbb{R}$, θ *a Cartan involution,* $\mathfrak{g} = \mathfrak{k} + \mathfrak{p}$ *the corresponding Cartan decomposition. Let* G *be any connected Lie group having* $\mathfrak{g}$ *as its Lie algebra. Denote by* K *the analytic subgroup of* G *having Lie algebra* $\mathfrak{k}$. *Then:*

(i) K *is closed and contains* $Z = \text{Cent } G$.

(ii) *There is a unique involutive automorphism* $\tilde{\theta}$ *of* G *whose differential is* θ.

(iii) *The map* $(k,X) \to k \exp X$, $K \times \mathfrak{p} \to G$ *is an analytic diffeomorphism of* $K \times \mathfrak{p}$ *onto* G.

(iv) K *is compact* $\Longleftrightarrow$ Z *is finite.*

(v) *If* Z *is finite,* K *is a maximal compact subgroup of* G *and any two such are conjugate.*

EXAMPLES. (1) In Exercise 2, p. 5, each compact group there is a maximal compact subgroup of its corresponding complexification.

(2) If $G = SL(n,\mathbb{R})$, then $K = SO(n)$.

(3) If $G = SO(m,n) = \{g \in SL(m+n,R):\ g$ preserves a quadratic form of signature $(m,n)\}$, then its neutral component G^0 has a maximal compact $K \cong SO(m) \times SO(n)$, whenever $m \geqq n \geqq 1$, $m > 1$ and $m+n \neq 4$.

For more examples, see Helgason [1].

We remark that $\#(Z) = \infty$ is a possibility. Indeed the universal covering group of $SL(2,\mathbb{R})$ has infinite center. It's well known that if G has a faithful matrix representation, it must have finite center, but the converse is false (take a finite covering of $SL(2,\mathbb{R})$). Also complex semisimple groups always have finite center.

Our next goal is the so-called Iwasawa decomposition. Let $\mathfrak{g}$ be a semisimple Lie algebra over $\mathbb{R}$, $\mathfrak{g} = \mathfrak{k} + \mathfrak{p}$ a Cartan decomposition. Let $\mathfrak{a}$ be a maximal subalgebra of $\mathfrak{g}$ contained in $\mathfrak{p}$. Since $[\mathfrak{a},\mathfrak{a}] \subseteq \mathfrak{p}$ and $[\mathfrak{a},\mathfrak{a}] \subseteq [\mathfrak{p},\mathfrak{p}] \subseteq \mathfrak{k}$, $\mathfrak{a}$ is automatically abelian. Then $\text{ad}_{\mathfrak{g}}(\mathfrak{a})$ forms a commutative family of semisimple automorphisms of $\mathfrak{g}$ (because they are symmetric and hence diagonalizable over $\mathbb{R}$). For $\lambda \in \text{Hom}_R(\mathfrak{a},\mathbb{R})$, we set $\mathfrak{g}_\lambda = \{X \in \mathfrak{g} : \text{ad } H(X) = \lambda(H)X,\ \forall\, H \in \mathfrak{a}\}$. We say that λ is a *restricted root* whenever

$\lambda \neq 0$ and $\mathfrak{g}_\lambda \neq \{0\}$. The simultaneous diagonalization guarantees that $\mathfrak{g} = \sum_\lambda \mathfrak{g}_\lambda$. As before $[\mathfrak{g}_\lambda, \mathfrak{g}_\mu] \subseteq \mathfrak{g}_{\lambda+\mu}$ and $B_\theta(\mathfrak{g}_\lambda, \mathfrak{g}_\mu) = 0$ if $\lambda \neq \mu$ (since $\theta \mathfrak{g}_\lambda = \mathfrak{g}_{-\lambda}$).

Let $\mathfrak{g}_c$ be the complexification of $\mathfrak{g}$. Put $\mathfrak{u} = \mathfrak{k} + i\mathfrak{p}$. It's easy to see that $\mathfrak{u}$ is a compact real form of $\mathfrak{g}_c$. Now extend $\mathfrak{a}$ to a maximal abelian subalgebra $\mathfrak{h}$ of $\mathfrak{g}$. Then $\mathfrak{h}$ is automatically θ-invariant. Indeed, if $X \in \mathfrak{h}$, $Y \in \mathfrak{a}$, then

$$[X - \theta X, Y] = [X,Y] - \theta[X,\theta Y] = [X,Y] + \theta[X,Y] = 0.$$

Since $X - \theta X \in \mathfrak{p}$ it follows from the maximality that $X - \theta X \in \mathfrak{a} \Rightarrow \theta X \in \mathfrak{h}$. Hence $\mathfrak{h} = (\mathfrak{h} \cap \mathfrak{k}) + (\mathfrak{h} \cap \mathfrak{p}) = (\mathfrak{h} \cap \mathfrak{k}) + \mathfrak{a}$. It is now a simple matter to verify that $\mathfrak{h}_c$ is a Cartan subalgebra of $\mathfrak{g}_c$. If $\mathfrak{g}_c = \mathfrak{h}_c + \sum_{\alpha \in \Delta} \mathfrak{g}_c^\alpha$, then clearly $\mathfrak{g}_\lambda = \left(\sum_{\substack{\alpha \in \Delta \\ \alpha|_{\mathfrak{a}} = \lambda}} \mathfrak{g}^\alpha \right) \cap \mathfrak{g}$.

PROPOSITION 9. *Let* $\mathfrak{m}$ *be the centralizer of* $\mathfrak{a}$ *in* $\mathfrak{k}$. *Then* $\mathfrak{g}_0 = \mathfrak{a} + \mathfrak{m}$ *and* $\mathfrak{m}_c = (\mathfrak{h} \cap \mathfrak{k})_c + \sum_{\substack{\alpha \in \Delta \\ \alpha|_{\mathfrak{a}} = 0}} \mathfrak{g}_c^\alpha$.

It is clear that $\mathfrak{a} + \mathfrak{m} \subseteq \mathfrak{g}_0$. Conversely, since $\mathfrak{g}_0$ is θ-invariant it must be that $\mathfrak{g}_0 = (\mathfrak{g}_0 \cap \mathfrak{k}) + (\mathfrak{g}_0 \cap \mathfrak{p}) = \mathfrak{m} + \mathfrak{a}$. The latter statement is proven by similar reasoning.

REMARK. It is obvious that $\mathfrak{a} \subseteq \mathfrak{h}_c^*$ (notation of Theorem 3). Hence it is possible to choose compatible orderings on the vector spaces $\mathfrak{a} \subseteq \mathfrak{h}_c^*$. We assume that is done henceforth. We always write Δ, Σ for the roots and restricted roots respectively. Given the order, we write Δ^+ (or Σ^+) for positive (restricted) roots and Δ^- (or Σ^-) for negative (restricted) roots.

DEFINITION. Set $\mathfrak{n} = \sum_{\lambda \in \Sigma^+} \mathfrak{g}_\lambda$.

$\mathfrak{n}$ is a subalgebra of $\mathfrak{g}$ and we have a vector space direct sum $\mathfrak{g} = \mathfrak{m} + \mathfrak{a} + \mathfrak{n} + \theta\mathfrak{n}$.

THEOREM 10. $\mathfrak{n}$ *is nilpotent,* $\mathfrak{b} = \mathfrak{a} + \mathfrak{n}$ *is solvable, and we have the direct sum decomposition* $\mathfrak{g} = \mathfrak{k} + \mathfrak{a} + \mathfrak{n}$.

Proof. Since Σ^+ is finite and $[\mathfrak{g}_\lambda, \mathfrak{g}_\mu] \subseteq \mathfrak{g}_{\lambda+\mu}$, the nilpotence of $\mathfrak{n}$ is clear. Also $[\mathfrak{b}, \mathfrak{b}] = [\mathfrak{a} + \mathfrak{n}, \mathfrak{a} + \mathfrak{n}] \subseteq [\mathfrak{a}, \mathfrak{n}] + \mathfrak{n} \subseteq \mathfrak{n} \Rightarrow \mathfrak{b}$ is solvable. The sum is direct because if $X \in \mathfrak{k}$, $H \in \mathfrak{a}$, $Y \in \mathfrak{n}$ is such that $X + H + Y = 0$, then $X - H + \theta Y = 0 \Rightarrow 2H + Y - \theta Y = 0$. Since the sum $\mathfrak{g} = \mathfrak{m} + \mathfrak{a} + \mathfrak{n} + \theta\mathfrak{n}$ is direct, we conclude that $H = 0$ and $Y = 0$. Therefore $X = 0$ as well. Finally $\mathfrak{g} = \mathfrak{m} + \mathfrak{a} + \mathfrak{n} + \theta\mathfrak{n} = \mathfrak{k} + \mathfrak{a} + \mathfrak{n} + \theta\mathfrak{n}$. But $Y \in \theta\mathfrak{n} \Rightarrow Y = (Y + \theta Y) - \theta Y \in \mathfrak{k} + \mathfrak{n}$.

Next we want to get the corresponding decomposition on groups. For that, we shall use the following two facts:

(1) There exists a basis of $\mathfrak{g}_c$ such that the matrices of the endomorphisms in ad $\mathfrak{g}_c$ have the following properties: (a) those of $\mathrm{ad}(\mathfrak{u})$ are skew hermitian, (b) those of $\mathrm{ad}(\mathfrak{a})$ are real diagonal, and (c) those of $\mathrm{ad}(\mathfrak{n})$ are upper triangular with zeroes on the diagonal.

(2) Let U be a Lie group with Lie algebra $\mathfrak{u}$, and subalgebras $\mathfrak{u}_1, \mathfrak{u}_2$ such that $\mathfrak{u} = \mathfrak{u}_1 \oplus \mathfrak{u}_2$. Then, if U_1, U_2 are the corresponding analytic subgroups the map $U_1 \times U_2 \to U$, $(x,y) \to xy$, is regular.

THEOREM 11. (Iwasawa Decomposition) *Let* G *be a connected semisimple Lie group with* $\mathfrak{g}$ *its Lie algebra, and* $\mathfrak{g} = \mathfrak{k} + \mathfrak{a} + \mathfrak{n}$ *as in* Theorem 10. *Let* K, A, N *be the corresponding analytic subgroups of* G. *Then the map* $(k,a,n) \to kan$ *is an analytic diffeomorphism of* $K \times A \times N$ *onto* G.

Proof. (Sketch) By a simple covering argument (which we don't include), matters are reduced to the case $G = \mathrm{Ad}(\mathfrak{g})$. Then the elements of G are endomorphisms of $\mathfrak{g}$ (and so of $\mathfrak{g}_c$) which in matrix terms have the following properties: those in K must be unitary matrices, elements of A give diagonal matrices with positive entries, and elements in N become upper triangular with 1 on the diagonal (all by (1) above). From this it is clear that the map is injective. A is clearly a simply connected closed subgroup (in fact a vector group) in G and N is a simply connected closed nilpotent subgroup of G. It follows that $B = AN$ is a simply connected closed subgroup of G having Lie algebra $\mathfrak{b} = \mathfrak{a} + \mathfrak{n}$. Finally one proves that $K \times B \to G$, $(k,b) \to kb$ is onto and then the last statement follows from (2) above.

EXAMPLES. (1) $G = SL(n,\mathbb{C})$, $K = SU(n)$, $A = \left\{ \begin{pmatrix} a_1 & & 0 \\ & \ddots & \\ 0 & & a_n \end{pmatrix} : a_i > 0, a_1 \cdots a_n = 1 \right\}$, $N = \left\{ \begin{pmatrix} 1 & & n_{ij} \\ & \ddots & \\ 0 & & 1 \end{pmatrix},\ n_{ij} \in \mathbb{C} \right\}$.

(2) $G = SL(n,\mathbb{R})$, $K = SO(n)$, A as above, N as above with real coefficients. Let's be more explicit in case $G = SL(2,\mathbb{R})$. If $g = \begin{pmatrix} \alpha & \beta \\ \gamma & \delta \end{pmatrix}$, $\alpha\delta - \beta\gamma = 1$, then $g = kan$ where

$$k = \begin{pmatrix} \frac{\alpha}{\varepsilon} & -\frac{\gamma}{\varepsilon} \\ \frac{\gamma}{\varepsilon} & \frac{\alpha}{\varepsilon} \end{pmatrix},\quad \varepsilon = \sqrt{\alpha^2 + \gamma^2},\quad a = \begin{pmatrix} \varepsilon & 0 \\ 0 & \varepsilon^{-1} \end{pmatrix},\quad n = \begin{pmatrix} 1 & x \\ 0 & 1 \end{pmatrix},\quad x = \begin{cases} \dfrac{\beta\varepsilon^2+\gamma}{\alpha\varepsilon^2} & \alpha \neq 0 \\[2ex] \dfrac{\delta\varepsilon^2-\alpha}{\gamma\varepsilon^2} & \gamma \neq 0. \end{cases}$$

(3) As the preceding computations reveal, even in very low-dimensional cases, the actual formulas that describe an Iwasawa decomposition may be quite complicated. However by being a little less explicit, one can obtain some useful information. For example, consider $G = Sp(2,\mathbb{C})$. It is well known (see e.g. Humphreys [1]) that $\mathfrak{g} = LA(G)$ has four positive roots of the form α, β, $\alpha + \beta$, $2\alpha + \beta$. Therefore it is easy to see that in an Iwasawa decomposition $G = KAN$, we must have $K \cong Sp(2)$, $A \cong (R_+^*)^2$, and N is a complex simply connected nilpotent Lie group whose Lie algebra has generators W, X, Y, Z

satisfying $[W,X] = Y$, $[W,Y] = Z$. In Chapter IV we shall compute the representation theory of the corresponding real form of N.

Next we want to discuss the Weyl group and the Bruhat decomposition. So let G be a connected semisimple Lie group, $\mathfrak{g}$ its Lie algebra, $\mathfrak{g} = \mathfrak{k} + \mathfrak{p}$ a Cartan decomposition, and $\mathfrak{a} \subseteq \mathfrak{p}$ a maximal abelian subalgebra. Let K be the Lie subgroup of G corresponding to $\mathfrak{k}$. Then $\mathrm{Ad}_G(k)$, $k \in K$, leaves $\mathfrak{p}$ invariant, but not $\mathfrak{a}$ of course.

DEFINITION. Set $M = Z(\mathfrak{a}) \cap K$ = centralizer of $\mathfrak{a}$ in $K = \{k \in K\colon \mathrm{Ad}(k)H = H \ \ \forall\ H \in \mathfrak{a}\}$, $M' = N(\mathfrak{a}) \cap K$ = normalizer of $\mathfrak{a}$ in $K = \{k \in K\colon \mathrm{Ad}(k)\mathfrak{a} \subseteq \mathfrak{a}\}$.

Clearly M is a normal subgroup of M'. But neither need be connected.

LEMMA 12. *M and M' both have $\mathfrak{m}$ as their Lie algebra.*

Proof. That $\mathfrak{m}$ is the Lie algebra of M is obvious. Next, suppose $Y \in LA(M')$. Then $[Y,\mathfrak{a}] \subseteq \mathfrak{a}$. Therefore for all $H,H' \in \mathfrak{a}$, $[[Y,H'],H] = 0$. But by the invariance of the Killing form B under ad, we have

$$B([Y,H],[Y,H]) = -B([H,[H,Y]],Y) = 0.$$

However B is strictly positive-definite on $\mathfrak{a}$. Therefore $[Y,H] = 0 \Longrightarrow Y \in \mathfrak{m}$.

Now $M'/M \subseteq K/M \subseteq (K/Z)/(M/Z)$ is a compact group. Hence M'/M is a compact Lie group with trivial Lie algebra. Therefore M'/M is a finite group. We write $W = M'/M$ and call W the *Weyl group*.

Let us denote $\mathfrak{a}' = \{H \in \mathfrak{a} : \lambda(H) \neq 0 \ \ \forall\ \lambda \in \Sigma\}$. By a *Weyl chamber* of $\mathfrak{a}$ we mean a connected component of $\mathfrak{a}'$. It is clear that

W acts on $\mathfrak{a}$. For each $\lambda \in \Sigma$ define $H_\lambda \in \mathfrak{a}$ as usual via $B(H,H_\lambda) = \lambda(H)$, $H \in \mathfrak{a}$. Then the function

$$s_\lambda : H \to H - \frac{2\lambda(H)}{\lambda(H_\lambda)} H_\lambda$$

defines a reflection in $\mathfrak{a}$ through the hyperplane perpendicular to H_λ.

THEOREM 13. (i) *The group* W *is generated by the reflections* s_λ, $\lambda \in \Sigma$.

(ii) W *permutes the Weyl chambers in a simply transitive fashion.*

Note also that W acts as a transformation group on A, $s(\exp H) = \exp sH$, $s \in W$. Of course then it also acts on $\hat{A}$. W does not act on M, but it will be important for us later to note that W acts on $\hat{M}$. The action is $(s\cdot\sigma)(m) = \sigma(x^{-1}mx)$, $m \in M$, $s = xM \in W$, $\sigma \in \hat{M}$.

We conclude this section with a nearly complete proof of the famous Bruhat decomposition. The details are taken from the notes of a course given by Helgason at MIT in 1966. It is based on the proof in Harish-Chandra [2].

First we change the notation slightly. Write $B = MAN$. This is a group because M normalizes $\mathfrak{n}$. Indeed for $H \in \mathfrak{a}$, $X \in \mathfrak{g}_\lambda$, $m \in M$, we have $[H,\mathrm{Ad}(m)X] = \mathrm{Ad}(m)[H,X] = \alpha(H)\mathrm{Ad}(m)X$. Thus M actually normalizes each restricted root space.

THEOREM 14. (Bruhat Decomposition) *The double coset space* $B\backslash G/B$ *is finite. In fact if* m_i', $1 \leq i \leq w$, *is a set of representatives for* $W = M'/M$, *then* $G = \bigcup_{i=1}^{w} Bm_i'B$, $w = \#(W)$.

We will need three lemmas.

LEMMA 15. *Let* $H \in \mathfrak{a}'$. *The map* $\phi: n \to \mathrm{Ad}_G(n)H - H$ *is a bijection of* N *onto* $\mathfrak{n}$.

Proof. $\mathrm{Ad}(\exp H)H - H = e^{\mathrm{ad}\,X}(H) - H = [X,H] + \frac{1}{2}[X,[X,H]] + \ldots \in \mathfrak{n}$ if $X \in \mathfrak{n}$. Therefore $\mathrm{Ad}(n)H - H \in \mathfrak{n}$.

ϕ is 1-1. Suppose $\mathrm{Ad}(n_1)H - H = \mathrm{Ad}(n_2)H - H$. Then $\mathrm{Ad}(n_1^{-1}n_2)H = H$. Let $n_1^{-1}n_2 = \exp X$, $X \in \mathfrak{n}$. Write $X = \sum_{\lambda \in \Sigma^+} X_\lambda$, $X_\lambda \in \mathfrak{g}_\lambda$. Then

$$\mathrm{Ad}(\exp X)H = H \Rightarrow [X,H] + \tfrac{1}{2}[X,[X,H]] + \ldots = 0.$$

Taking a smallest λ such that $X_\lambda \neq 0$ we see that $\lambda(H)X_\lambda = 0$. Thus $\lambda(H) = 0$, which contradicts the regularity of H. Hence $n_1 = n_2$.

ϕ is onto. If not, take $Z \in \mathfrak{n} - \phi(N)$, and expand $Z = \sum_{\lambda \in \Sigma^+} Z_\lambda$. Not all Z_λ are zero. Let λ_0 be the smallest such one. Select $Z \in \mathfrak{n} - \phi(N)$ such that λ_0 is as large as possible. Since H is regular we can take $Z_1 \in \mathfrak{n}$ so that $\mathrm{ad}\,H(Z_1) = Z$. Set $n_1 = \exp Z_1$. Then

$$\begin{aligned}\mathrm{Ad}(n_1)(H+Z) - H &\equiv [Z_1,H] + [Z_1,[Z_1,H]] + \ldots + Z \mod \sum_{\lambda > \lambda_0} \mathfrak{g}_\lambda \\ &\equiv [Z_1,H] + Z \\ &\equiv 0.\end{aligned}$$

Therefore by the choice of Z, there is $n' \in N$ such that $\mathrm{Ad}(n')H - H = \mathrm{Ad}(n_1)(H+Z) - H$. That is $\mathrm{Ad}(n_1^{-1}n)H - H = Z$, which is a contradiction.

Next let $\mathfrak{b} = \mathrm{LA}(B) = \mathfrak{m} + \mathfrak{a} + \mathfrak{n}$. For $x \in G$, set $\mathfrak{b}_x = \mathfrak{b} \cap \mathrm{Ad}(x)\mathfrak{b}$.

LEMMA 16. *For any* $x \in G$, *we have* $\mathfrak{b} = \mathfrak{b}_x + \mathfrak{n}$.

Proof. The inclusion $\mathfrak{b} \supseteq \mathfrak{b}_x + \mathfrak{n}$ is obvious. Thus it suffices to prove $\dim(\mathfrak{b}_x + \mathfrak{n}) = \dim \mathfrak{b}$. But $\dim(\mathfrak{b}_x + \mathfrak{n}) = \dim \mathfrak{b}_x + \dim \mathfrak{n} - \dim(\mathfrak{b}_x \cap \mathfrak{n})$. Moreover $\mathfrak{n} \cap \mathfrak{b}_x = \mathfrak{n} \cap \mathrm{Ad}(x)\mathfrak{b} = \mathfrak{n} \cap \mathrm{Ad}(x)\mathfrak{n}$ (because $\mathfrak{n} = \{X \in \mathfrak{b} : \mathrm{ad}\, X \text{ is nilpotent}\}$ -- see Lemma 17). Therefore if $x = kan$, then $\mathrm{Ad}(x)\mathfrak{n} = \mathrm{Ad}(k)\mathfrak{n}$, and we see it is no loss of generality to assume $x \in K$.

Now $\mathfrak{g} = \sum \mathfrak{g}_\lambda$ and this decomposition is orthogonal with respect to B_θ. But since $\mathfrak{b} = \sum\limits_{\lambda \geqq 0} \mathfrak{g}_\lambda$, $\mathfrak{n} = \sum\limits_{\lambda > 0} \mathfrak{g}_\lambda$, and $\theta \mathfrak{g}_\lambda = \mathfrak{g}_{-\lambda}$, we have $\mathfrak{b}^\perp = \theta\mathfrak{n}$. Also since B_θ is invariant under $\mathrm{Ad}(x)$, $x \in K$, one obtains easily that $(\mathrm{Ad}(x)\mathfrak{b})^\perp = \mathrm{Ad}(x)\theta\mathfrak{n}$. Therefore $(\mathfrak{b} + \mathrm{Ad}(x)\mathfrak{b})^\perp = \mathfrak{b}^\perp \cap (\mathrm{Ad}(x)\mathfrak{b})^\perp = \theta\mathfrak{n} \cap \mathrm{Ad}(x)\theta\mathfrak{n} = \theta(\mathfrak{n} \cap \mathrm{Ad}(x)\mathfrak{n})$. Thus we have shown

$$\begin{aligned}\dim(\mathfrak{b}_x + \mathfrak{n}) &= \dim \mathfrak{b}_x + \dim \mathfrak{n} - (\dim \mathfrak{g} - \dim(\mathfrak{b} + \mathrm{Ad}(x)\mathfrak{b}))\\ &= \dim \mathfrak{b}_x + \dim \mathfrak{n} - (\dim \mathfrak{g} - (2\dim \mathfrak{b} - \dim \mathfrak{b}_x))\\ &= \dim \mathfrak{n} - \dim \mathfrak{g} + 2\dim \mathfrak{b}\\ &= \dim \mathfrak{b}.\end{aligned}$$

We now state the third lemma, but we omit full details of the proof.

LEMMA 17. (i) *$\mathfrak{n}$ is the set of elements Z in $\mathfrak{b}$ such that* $\mathrm{ad}\, Z$ *is nilpotent.*

(ii) *$\mathfrak{a} + \mathfrak{n}$ is the set of elements Z in $\mathfrak{b}$ such that all eigenvalues of* $\mathrm{ad}\, Z$ *in $\mathfrak{g}_c$ are real.*

Method of proof. One needs to embed $\mathfrak{a} \subseteq \mathfrak{h} \subseteq \mathfrak{h}_c$ and look at the various root spaces to show that the eigenvalues corresponding to $\mathfrak{m}$ are imaginary.

Finally we come to the

Proof of Theorem 14. Let $x \in G$ and $H \in \mathfrak{a}'$. By Lemma 16, there is $X \in \mathfrak{n}$ such that $H + X \in \mathrm{Ad}(x)\mathfrak{b} \cap \mathfrak{b}$. By Lemma 15, there is $n_1 \in N$ such that $\mathrm{Ad}(n_1)H - H = X$. Therefore $\mathrm{Ad}(x^{-1})\mathrm{Ad}(n_1)H \in \mathfrak{b}$. Write $\mathrm{Ad}(x^{-1}n_1)H = T' + H' + X'$, $T' \in \mathfrak{m}$, $H' \in \mathfrak{a}$, $X' \in \mathfrak{n}$. It follows from Lemma 17 that $T' = 0$. Also $H' \in \mathfrak{a}'$. Therefore using Lemma 15 once again, we can find $n_2 \in N$ such that $\mathrm{Ad}(n_2)H' - H' = X'$. Combining results, we get $\mathrm{Ad}(n_2^{-1}\, x\, n_1^{-1})H = H'$. But the centralizer of $H_1 \in \mathfrak{a}'$ in $\mathfrak{g}$ is just $\mathfrak{m} + \mathfrak{a}$. Therefore $\mathrm{Ad}(n_2^{-1}\, x\, n_1^{-1})(\mathfrak{m} + \mathfrak{a}) = \mathfrak{m} + \mathfrak{a}$. Again by Lemma 17, we conclude $\mathrm{Ad}(n_2^{-1}\, x\, n_1^{-1})\mathfrak{a} = \mathfrak{a}$. However the normalizer of $\mathfrak{a}$ in G is $M'A$. Thus $n_2^{-1}\, x\, n_1^{-1} \in M'A \Longrightarrow$ $x \in n_2 M' A n_1 \subseteq \bigcup B m_i' B$.

It remains to prove disjointness. Suppose $b_1 m_i' = m_j' b_2$. Choose $H \in \mathfrak{a}'$ such that $\mathrm{Ad}(m_i')H \in \mathfrak{a}^+$. Set $H_i = \mathrm{Ad}(m_i')H$. Then

$$\mathrm{Ad}(b_1)H_i = \mathrm{Ad}(m_j')\mathrm{Ad}(b_2)H = \mathrm{Ad}(m_j')\mathrm{Ad}(n_2)H.$$

Now

$$\mathrm{Ad}(n_2)H = H + [X_2,H] + \tfrac{1}{2}[X_2,[X_2,H]] + \dots$$

$$\mathrm{Ad}(b_1)H_i = H_i + \dots .$$

By comparing root spaces we find $H_i = \mathrm{Ad}(m_j')H = \mathrm{Ad}(m_j' m_i^{-1})H_i$. Then using Theorem 13 (ii), we conclude that $\mathrm{Ad}(m_j' m_i^{-1})|_{\mathfrak{a}} = 1 \Longrightarrow m_j' m_i^{-1} \in M$.

EXAMPLES. (1) $G = SL(n,\mathbb{R})$, $B = \left\{\begin{pmatrix} a_1 & & x_{ij} \\ & \ddots & \\ 0 & & a_n \end{pmatrix} : a_1 \cdots a_n = 1,\ x_{ij} \in \mathbb{R}\right\}$, $W = S_n$ the symmetric group on n letters. The Bruhat decomposition reflects the well-known fact that every $g \in G$ can be written $g = b_1 t b_2$, $b_1, b_2 \in B$, t a permutation matrix of determinant 1.

(2) $G = Sp(2,\mathbb{C})$, $B = TN$ where N is the group of Example 3, p. 10, and $T \cong (\mathbb{C}^*)^2$. Also W is isomorphic to a semidirect product of $\mathbb{Z}_2^2$ with S_2 -- the latter acting on the former in the obvious manner.

B. VARIOUS SERIES OF REPRESENTATIONS

In this section we shall study the different types of representations of semisimple Lie groups that have been discovered over the years. The first examples were created in an ad hoc manner by the Russian school (led by Gelfand). Later Harish-Chandra and others put the material into a more systematic framework. It is customary to group the representations into various "series". We follow that practice here.

1. Principal:minimal parabolic. Let G be a connected semisimple Lie group with finite center. Let $G = KAN$ be an Iwasawa decomposition of G, M = the centralizer of A in K. The group $B = MAN$ is called a *minimal parabolic* subgroup of G (see no. 3 for a general discussion of parabolic groups). We first set down some representations of B. Let $\sigma \in \hat{M}$, $\tau \in \hat{A}$ and form the finite-dimensional irreducible unitary representation $\sigma \times \tau$ of B given by $(\sigma \times \tau)(man) = \sigma(m)\tau(a)$.

EXERCISE. Using Lie's Theorem (AN is solvable) and the fact that $[\mathfrak{a} + \mathfrak{n}, \mathfrak{a} + \mathfrak{n}] = \mathfrak{n}$, show that the representations $\sigma \times \tau$ exhaust the finite-dimensional members of $\hat{B}$.

The *principal series* of representations corresponding to the minimal parabolic subgroup B constitute the family of representations

$$\pi(\sigma,\tau) = \mathrm{Ind}_B^G\ \sigma \times \tau.$$

Let us, using the procedures outlined in Appendix A, write out these representations in more detail.

First of all G is unimodular, but B is not. In fact, one computes fairly easily that $\Delta_B(man) = e^{2\rho(H)}$, $\exp H = a \in A$, where $\rho = \frac{1}{2} \sum_{\lambda \in \Sigma^+} (\dim \mathfrak{g}_\lambda)\lambda$. Therefore from the decomposition $G = ANK$ (note: $KAN = G = G^{-1} = NAK = ANK$), we see that the q function may be defined by $q_B(ank) = e^{2\rho(H)}$, $\exp H = a$. Then the space $\mathcal{H}_{\pi(\sigma,\tau)}$ of $\pi(\sigma,\tau)$ is given by

$$\mathcal{H}_{\pi(\sigma,\tau)} = \Big\{f: G \to \mathcal{H}_\sigma, \ f \text{ measurable},$$

$$f(mang) = \tau(a)\sigma(m)f(g), \quad man \in B, \quad a.a.\ g \in G,$$

$$\int_{G/B} \|f(\bar{g})\|^2 d\bar{g} < \infty\Big\}$$

where as usual $d\bar{g}$ denotes the quasi-invariant measure determined by q_B. The action of $\pi(\sigma,\tau)$ is as follows

$$\pi(\sigma,\tau)(g)f(x) = f(xg)[q_B(xg)/q_B(x)]^{\frac{1}{2}}, \quad x, g \in G.$$

EXERCISE. Apply Exercise (4) of Appendix A to see that the representations $\pi(\sigma,\tau)$ can also be realized on the space

$$\mathcal{H}'_{\pi(\sigma,\tau)} = \Big\{f: G \to \mathcal{H}_\sigma, \ f \text{ measurable},$$

$$f(mang) = q_B^{\frac{1}{2}}(a)\tau(a)\sigma(m)f(g), \quad man \in B, \quad a.a.\ g \in G,$$

$$\sup_{\phi \in D} \int_G \|f(g)\|^2 \phi(g)dg < \infty\Big\},$$

where $D = \{\phi \in C_0^+(G): \int_B \phi(bg)db \leqq 1, \quad g \in G\}$. On this space the action of the group becomes

$$\pi(\sigma,\tau)(g)f(x) = f(xg).$$

When computing it is often convenient to have still other realizations of these representations. We give two more -- the so-called "compact" and "nilpotent" realizations. First, the compact one. Let

$$L_2(K;M;\mathcal{H}_\sigma) = \Big\{f: K \to \mathcal{H}_\sigma, \ f \text{ measurable},$$

$$f(mk) = \sigma(m)f(k), \quad m \in M, \quad a.a.\ k \in K,$$

$$\int_{K/M} \|f(\bar{k})\|^2 d\bar{k} < \infty\Big\}.$$

$d\bar{k}$ of course denotes the K-invariant measure on K/M. The mapping $f \to f|_K$, $\mathcal{H}'_{\pi(\sigma,\tau)} \to L_2(K;M;\mathcal{H}_\sigma)$ is a unitary mapping of Hilbert spaces. The inverse mapping is given by $F \to f$ where

$$f(ank) = q_B^{\frac{1}{2}}(a)\tau(a)F(k).$$

Transferring the action of $\pi(\sigma,\tau)$ to the space $L_2(K;M;\mathcal{H}_\sigma)$ we find the formula

$$\pi(\sigma,\tau)(g)F(k_1) = q_B^{\frac{1}{2}}(a(k_1g))\tau(a(k_1g))F(k(k_1g))$$

where for $x \in G$ we have written $x = a(x)n(x)k(x)$.

Next we give the nilpotent realization. For that we need to comment on a further refinement of the Bruhat decomposition $G = \bigcup Bm_i'B$. It is known (Moore [1]) that each double coset is a manifold of lower dimension in G with the exception of one. That one is obtained as follows. The Weyl group W acts on A and $\mathfrak{a}$. Let $s \in W$, $s = m'M$, $m' \in M'$ and take a restricted root $\lambda \in \Sigma$, $X \in \mathfrak{g}_\lambda$, $H \in \mathfrak{a}$. Then

$$\lambda(H)\mathrm{Ad}\,m'(X) = \mathrm{Ad}\,m'[H,X] = [\mathrm{Ad}\,m'(H)\,,\,\mathrm{Ad}\,m'(X)] = [sH\,,\,\mathrm{Ad}\,m'(X)].$$

Replace sH by H to get

$$[H\,,\,\mathrm{Ad}\,m'(X)] = s^{-1}\lambda(H)\mathrm{Ad}\,m'(X).$$

This shows that if $\lambda \in \Sigma$, then $s\lambda \in \Sigma$ as well.

LEMMA 1. *There exists a unique element* $s_0 \in W$ *such that for all* $\lambda \in \Sigma^+$, *we have* $s_0\lambda \in \Sigma^-$.

The double coset $Bm_0'B$, $s_0 = m_0'M$, is the unique coset whose dimension equals that of G. It follows that up to a set of measure zero $G = Bm_0'B = Bm_0'MAN = Bm_0'N = Bm_0'Nm_0'^{-1}m_0' = BVm_0'$, $V = \tilde{\theta}N = \exp\theta\mathfrak{n}$. Thus we see that the manifold BV differs from G by a set of measure zero.

For our nilpotent realization we take

$$L_2(V;\mathcal{H}_\sigma) = \{f\colon V \to \mathcal{H}_\sigma,\ \ f \text{ measurable},\ \int_V \|f(v)\|^2 dv < \infty\}.$$

The mapping $f \to f|_V$, $\mathcal{H}'_{\pi(\sigma,\tau)} \to L_2(V;\mathcal{H}_\sigma)$ is a unitary mapping of Hilbert spaces. The inverse mapping is given by $f \to \phi$

$$\phi(bv) = q_B^{\frac{1}{2}}(b)(\sigma \times \tau)(b)f(v).$$

The action of the representation becomes

$$\pi(\sigma,\tau)(g)\phi(v_1) = q_B^{\frac{1}{2}}(b(v_1g))(\sigma \times \tau)(b(v_1g))f(v(v_1g))$$

where for a.a. $x \in G$, we write $x = b(x)v(x) \in BV$.

EXAMPLE. Let $G = SL(2,\mathbb{C})$. Then $M = \left\{\begin{pmatrix} e^{it} & 0 \\ 0 & e^{-it}\end{pmatrix} : t \in \mathbb{R}\right\}$. The principal series is of the form $\pi(m,r) = \operatorname{Ind}_{MAN}^{G}(\sigma_m \times \tau_r)$, where

$$(\sigma_m \times \tau_r)\begin{pmatrix} \alpha & 0 \\ 0 & \alpha^{-1}\end{pmatrix} = |\alpha|^{ir}\left(\frac{\alpha}{|\alpha|}\right)^m, \quad r \in \mathbb{R},\quad m \in \mathbb{Z}.$$

Identifying $V = \left\{\begin{pmatrix} 1 & 0 \\ z & 1\end{pmatrix} : z \in \mathbb{C}\right\}$ with $\mathbb{C}$, we see that these representations act on $L_2(\mathbb{C})$ via

$$\pi(m,\tau)(g)f(z) = |bz+d|^{-2+ir}\left(\frac{bz+d}{|bz+d|}\right)^m f\left(\frac{az+c}{bz+d}\right), \quad g = \begin{pmatrix} a & b \\ c & d\end{pmatrix}.$$

EXERCISES. (1) Compute explicitly the compact realization of the principal series for $G = SL(2,\mathbb{C})$.

(2) Work out the compact and nilpotent realizations of the principal series for $G = SL(n,\mathbb{C})$, $n > 2$ and $G = SL(n,\mathbb{R})$ (see Gelfand and Naimark [1]).

For the sake of the reader's education, we elaborate at some length now on an important (and often illustrative) class of semi-simple Lie groups.

DEFINITION. The $\mathbb{R}$-rank of G is by definition the dimension of A.

Suppose in the following that the $\mathbb{R}$-rank of G is *one*. Since the Weyl group is generated by reflections, and since there is only one possible reflection, the Weyl group must have order two, $G = B \cup Bm_0'B$. Suppose in addition that G is simple (that is, $\mathfrak{g}$ is a simple Lie algebra). Then up to local isomorphism, it is possible to specify the group G. The possibilities are: $G = SO_e(n,1)$, $SU(n,1)$, $Sp(n,1)$ or the adjoint group of the exceptional Lie algebra $\mathfrak{f}_{4,9}$. Let us consider in detail the classical cases (namely the first three).

Let $n \geqq 2$ and set $\mathbb{K} = \mathbb{R}$, $\mathbb{C}$ or $\mathbb{H}$. Let G be the group of all automorphisms of $\mathbb{K}^{n+1}$ which preserve the hermitian quadratic form $x_1\bar{y}_1 + \dots + x_n\bar{y}_n - x_{n+1}\bar{y}_{n+1}$, and which (in case $\mathbb{K} = \mathbb{R}$ or $\mathbb{C}$) have determinant 1. Then G is $SO(n,1)$, $SU(n,1)$ or $Sp(n,1)$ according as $\mathbb{K} = \mathbb{R}$, $\mathbb{C}$ or $\mathbb{H}$. The latter two are connected and we denote by $SO_e(n,1)$ the neutral component of the former.

For the Lie algebra $\mathfrak{g}$ of G^0 one has a Cartan decomposition $\mathfrak{g} = \mathfrak{k} + \mathfrak{p}$ where

$$\mathfrak{k} = \begin{pmatrix} X_1 & 0 \\ 0 & X_2 \end{pmatrix}, \quad X_1 \text{ is } n \times n \text{ skew hermitian}, \quad X_2 \in \mathbb{K} \text{ is skew},$$

$$X_2 + \operatorname{tr} X_1 = 0 \quad (\mathbb{K} = \mathbb{C}), \qquad X_2 = 0 \quad (\mathbb{K} = \mathbb{R}),$$

$$\mathfrak{p} = \begin{pmatrix} 0 & Y \\ {}^t\bar{Y} & 0 \end{pmatrix}, \quad Y \text{ a column vector in } \mathbb{K}^n.$$

In each case the Cartan involution is negative conjugate transpose. If we make the choice

$$\mathfrak{a} = \mathbb{R}\begin{pmatrix} 0 & \mathrm{i} \\ \mathrm{i} & 0 \end{pmatrix}, \quad \mathrm{i} = \begin{pmatrix} 1 \\ 0 \\ \vdots \\ 0 \end{pmatrix}, \quad A = \begin{pmatrix} \cosh t & 0 & \sinh t \\ 0 & I & 0 \\ \sinh t & 0 & \cosh t \end{pmatrix},$$

then the positive restricted root spaces are

$$\mathfrak{g}_\lambda = \begin{pmatrix} 0 & {}^t\bar{X} & 0 \\ -X & 0 & X \\ 0 & {}^t\bar{X} & 0 \end{pmatrix}, \quad X \text{ a column vector in } \mathbb{K}^{n-1}$$

$$\mathfrak{g}_{2\lambda} = \begin{pmatrix} Y & 0 & -Y \\ 0 & 0 & 0 \\ Y & 0 & -Y \end{pmatrix}, \quad Y \in \mathbb{K}, \quad \bar{Y} = -Y,$$

where $\lambda\begin{pmatrix} 0 & \mathrm{i} \\ \mathrm{i} & 0 \end{pmatrix} = 1$. Furthermore $\mathfrak{n} = \mathfrak{g}_\lambda + \mathfrak{g}_{2\lambda}$, $\mathfrak{v} = \theta\mathfrak{n}$, and

$$N = \begin{pmatrix} 1 + Y - \frac{1}{2}|X|^2 & {}^t\bar{X} & -Y + \frac{1}{2}|X|^2 \\ -X & I & X \\ Y - \frac{1}{2}|X|^2 & {}^t\bar{X} & 1 - Y + \frac{1}{2}|X|^2 \end{pmatrix}$$

$$V = \begin{pmatrix} 1 + Y - \frac{1}{2}|X|^2 & {}^t\bar{X} & Y - \frac{1}{2}|X|^2 \\ -X & I & -X \\ -Y + \frac{1}{2}|X|^2 & -{}^t\bar{X} & 1 - Y + \frac{1}{2}|X|^2 \end{pmatrix}.$$

Also

$$K = \begin{cases} \begin{pmatrix} k & 0 \\ 0 & 1 \end{pmatrix} & k \in SO(n), \quad \mathbb{K} = \mathbb{R} \\ \begin{pmatrix} u & 0 \\ 0 & c \end{pmatrix} & u \in U(n), \quad c \in U(1), \quad (\det u)c = 1, \quad \mathbb{K} = \mathbb{C} \\ \begin{pmatrix} u & 0 \\ 0 & c \end{pmatrix} & u \in Sp(n), \quad c \in Sp(1), \quad \mathbb{K} = \mathbb{H}, \end{cases}$$

$$\mathfrak{m} = \begin{pmatrix} X_2 & 0 & 0 \\ 0 & X & 0 \\ 0 & 0 & X_2 \end{pmatrix}, \quad X_2 \text{ skew in } \mathbb{K}, \quad X \text{ an } (n-1)\times(n-1) \text{ skew hermitian},$$

$$2X_2 + \operatorname{tr} X = 0 \quad (\mathbb{K} = \mathbb{C}),$$

and

$$M = \begin{cases} \begin{pmatrix} 1 & 0 & 0 \\ 0 & u & 0 \\ 0 & 0 & 1 \end{pmatrix} & u \in SO(n-1), \quad \mathbb{K} = \mathbb{R} \\ \begin{pmatrix} c & 0 & 0 \\ 0 & \lambda & 0 \\ 0 & 0 & c \end{pmatrix} & c \in U(1), \quad \lambda \in U(n-1), \quad c^2 \det \lambda = 1, \quad \mathbb{K} = \mathbb{C} \\ \begin{pmatrix} c & 0 & 0 \\ 0 & \lambda & 0 \\ 0 & 0 & c \end{pmatrix} & c \in Sp(1), \quad \lambda \in Sp(n-1), \quad \mathbb{K} = \mathbb{H}. \end{cases}$$

Note that in case $\mathbb{K} = \mathbb{R}$, N is abelian whereas in the other two cases N is a two-step nilpotent group. Note also that M is connected here -- a fact that fails in general. We leave it to the reader to verify these details as well as to find a representative of the non-trivial coset in M'/M. The action of W on $\mathrm{Hom}_{\mathbb{R}}(\mathfrak{a},\mathbb{R})$ will of course be

$$\lambda \to s\lambda = \begin{cases} \lambda & s = e \\ -\lambda & s \neq e. \end{cases}$$

We also leave it to the reader to ponder the different realizations of the principal series representations $\pi(\sigma,\tau)$ for these groups.

We now state (without proof) the fundamental results on the behavior of these principal series. First recall the previous observation that W acts on both $\hat{A}$ and $\hat{M}$.

THEOREM 2. (i) (Bruhat [1]) $\mathscr{I}(\pi(\sigma_1,\tau_1),\pi(\sigma_2,\tau_2)) \leqq \#s \in W\colon s\sigma_1 \cong \sigma_2,\ s\tau_1 = \tau_2$. *In particular*

(a) $\pi(\sigma,\tau)$ *is a finite direct sum of irreducible representations.*

(b) $\pi(\sigma,\tau)$ *is irreducible if* $\forall\, s \in W,\ s \neq 1,\ s(\sigma\times\tau) \ncong \sigma\times\tau$. *For example, if* G *has* $\mathbb{R}$*-rank one, this is true for all pairs* $(\sigma,\tau) \in \hat{M}\times\hat{A},\ \tau \neq 1_A$.

(ii) (Bruhat [1]) $\pi(\sigma,\tau) \cong \pi(s\sigma,s\tau)$, $s \in W$.

(iii) (Kostant [1]) *If* $\sigma = 1_M$, *then* $\pi(1_M,\tau)$ *is irreducible* $\forall\ \tau \in \hat{A}$.

(iv) (Wallach [1]) *If* G *has only one conjugacy class of Cartan subgroups* (see no. 3 for the meaning of this phrase), *then* $\pi(\sigma,\tau)$ *is irreducible* $\forall\ \sigma,\tau$. *We mention that the condition is satisfied by complex groups and* $SO_e(2n+1,1)$.

Thus we see that "almost all" of the representations $\pi(\sigma,\tau)$ are irreducible. Actually, we are getting closer to prescribing exactly which are reducible and how they break up (see e.g. Knapp and Stein [1]). An interesting conjecture that is still not settled is: $\pi(\sigma,\tau) = \sum_i^{\oplus} \pi_i \Longrightarrow \pi_i \not\cong \pi_j$, $i \neq j$. (*Added after typing*: It was just settled in the affirmative by Knapp - see Bull. A.M.S, vol. 79, 1973.)

2. Discrete series. We begin with some general comments on square-integrable representations (which can be found in Dixmier [5]). For the moment we assume only that G is unimoduler.

DEFINITION. $\pi \in \hat{G}$ is called *square-integrable* if there is a non-zero vector $\xi \in \mathcal{H}_\pi$ such that $x \to (\pi(x)\xi,\xi)$ is a square-integrable function on G.

THEOREM 3. *Let* $\pi \in \hat{G}$ *be square-integrable. Then for every* $\xi,\eta \in \mathcal{H}_\pi$ *the function* $\phi_{\xi,\eta}(x) = (\pi(x)\xi,\eta)$, $x \in G$, *is square-integrable. Moreover there exists a constant* $d_\pi > 0$ *such that*

$$\int_G \phi_{\xi,\eta}(x)\bar{\phi}_{\xi',\eta'}(x)dx = d_\pi^{-1}(\xi,\xi')\overline{(\eta,\eta')}$$

$$\phi_{\xi,\eta} * \bar{\phi}_{\xi',\eta'} = d_\pi^{-1}(\xi,\eta')\phi_{\xi',\eta}.$$

The constant d_π is called the *formal dimension* of π.

THEOREM 4. *Let* $\pi_1, \pi_2 \in \hat{G}$ *be inequivalent square-integrable representations. Then*

$$\int_G \phi^1_{\xi,\eta}(x)\bar{\phi}^2_{\xi',\eta'}(x)dx = 0$$

$$\phi^1_{\xi,\eta} * \phi^2_{\xi',\eta'} = 0.$$

These are called the *orthogonality relations*. We shall write $\hat{G}_d$ to denote the subset of $\hat{G}$ consisting of square-integrable representations and call it the *discrete series*. The name formal dimension is motivated by the compact case where d_π is actually equal to $\dim \pi$.

EXERCISE. If $\hat{G}_d$ is not empty, show that $Z = \text{Cent } G$ must be compact.

Before considering the discrete series for semisimple groups, we need some further notions of rank. If K is a compact Lie group, $\mathfrak{k}$ its Lie algebra, then $\text{rank } K$ is the dimension of a maximal abelian subalgebra of $\mathfrak{k}$. If K is connected, it is also the dimension of a maximal torus T in K --, that is, a maximal connected abelian subgroup.

EXAMPLE. If $K = SO(n)$, then a maximal torus is given by

$$T = \begin{pmatrix} \cos\theta_1 & \sin\theta_1 & & & \\ -\sin\theta_1 & \cos\theta_1 & & & \\ & & \cos\theta_2 & \sin\theta_2 & \\ & & -\sin\theta_2 & \cos\theta_2 & \\ & & & & \ddots \end{pmatrix},$$

so that $\text{rank } SO(n) = \left[\frac{n}{2}\right]$.

If G is a connected semisimple Lie group, we define $\operatorname{rank} G = \operatorname{rank} M + \dim A$ = the dimension of a maximal abelian subalgebra of $\mathfrak{g}_0 = \mathfrak{m} + \mathfrak{a}$. (The reader is encouraged to see the next no. for another definition of rank G.)

LEMMA 5. (Harish-Chandra [8]) G *has a discrete series if and only if* $\operatorname{rank} G = \operatorname{rank} K$.

EXAMPLE. If $G = SO_e(n,1)$, then $\operatorname{rank} G = \operatorname{rank} SO(n-1) + 1 = \left[\frac{n-1}{2}\right] + 1$. Also $\operatorname{rank} K = \operatorname{rank} SO(n) = \left[\frac{n}{2}\right]$. These are equal exactly when n is even.

It is possible to give an explicit parameterization of the discrete series for a semisimple group G. To explain that we need some further notation. Let $\mathfrak{b} \subseteq \mathfrak{k} \subseteq \mathfrak{g}$, $\mathfrak{b}$ a maximal abelian subalgebra. Consider the complexifications $\mathfrak{b}_c \subseteq \mathfrak{k}_c \subseteq \mathfrak{g}_c$. It follows from the assumption $\operatorname{rank} K = \operatorname{rank} G$ that $\mathfrak{b}_c$ is a Cartan subalgebra of $\mathfrak{g}_c$. Let Q be a system of positive roots for $(\mathfrak{g}_c, \mathfrak{b}_c)$. Define the polynomial function ω on $\mathfrak{b}_c$ by

$$\omega(Y) = \prod_{\alpha \in Q} \alpha(Y).$$

Set $\mathcal{L} = \operatorname{Hom}_{\mathbb{C}}(\mathfrak{b}_c, \mathbb{C})$ and define $\mathcal{L}' = \{\lambda \in \mathcal{L} : \omega(H_\lambda) = \prod_{\alpha \in Q} B(H_\lambda, H_\alpha) \neq 0\}$. Next let B be the analytic subgroup of G corresponding to $\mathfrak{b}$. B is in fact a torus. Every $\chi \in \hat{B}$ determines a linear form $\lambda \in \operatorname{Hom}_{\mathbb{R}}(\mathfrak{b}, i\mathbb{R})$ by

$$\chi(\exp Y) = e^{\lambda(Y)}, \quad Y \in \mathfrak{b}.$$

The collection $\mathcal{F}$ of λ's thus obtained is a lattice, indeed

$$\mathcal{F} = \{\lambda \in \operatorname{Hom}_{\mathbb{R}}(\mathfrak{b}, i\mathbb{R}) : \lambda(Y) \in 2\pi i\mathbb{Z} \text{ whenever } \exp Y = e,\ Y \in \mathfrak{b}\}.$$

Finally set $W_B = N(B)/B$, a finite group which acts on $\mathcal{L}$ and leaves $\mathcal{L}'$ and $\mathcal{F}$ invariant.

THEOREM 6. (Harish-Chandra [8]) *There exists a* 1-1 *correspondence between equivalence classes of square-integrable irreducible unitary representations of* G *(the discrete series) and the set* $(\mathcal{L}' \cap \mathcal{F})/W_B$.

The correspondence is given explicitly in terms of the characters of the representations -- objects we will discuss in detail in section C. It is important to realize that Harish-Chandra does not write down the discrete series representations. In fact, a general realization of the discrete series is still not available, although there has been considerable progress on this problem. The most famous work in this area is that of Schmid [1].

EXAMPLE. Let $G = SL(2,\mathbb{R})$. In this case $M = \begin{pmatrix} \pm 1 & 0 \\ 0 & \pm 1 \end{pmatrix}$, rank G = dim A = rank K = 1. Then $B = K$, $\mathfrak{b} = \mathfrak{k} = \left\{\begin{pmatrix} 0 & b \\ -b & 0 \end{pmatrix} : b \in R\right\}$, $\mathcal{F}$ is a one-parameter lattice, and $\mathcal{L} - \mathcal{L}'$ contains only 0. Since $N(B) = B$, it must be that the discrete series is parameterized by a free abelian group on one generator without zero. In fact in this case, one knows how to write down the representations (see e.g. Sally [1]).

Let $n = 1, \frac{3}{2}, 2, \ldots$ and let $\mathcal{H}_n^+$ (resp. $\mathcal{H}_n^-$) denote the Hilbert space of holomorphic (resp. conjugate holomorphic) functions on the upper half plane P with inner product

$$(f,g)_n = \frac{1}{\Gamma(2n-1)} \int_P f(x+iy)\bar{g}(x+iy)y^{-2+2n} dx\, dy .$$

The space $\mathcal{H}_n^\pm$ is complete and the discrete series representations $\pi_n^\pm$ act via

$$\pi_n^+(g)f(z) = (bz+d)^{-2n} f\left(\frac{az+c}{bz+d}\right), \quad f \in \mathcal{H}_n^+$$

$$\pi_n^-(g)f(z) = (b\bar{z}+d)^{-2n} f\left(\frac{az+c}{bz+d}\right), \quad f \in \mathcal{H}_n^-,$$

$g = \begin{pmatrix} a & b \\ c & d \end{pmatrix} \in G.$

3. <u>Principal series : cuspidal parabolic</u>. In order to generalize the family of representations introduced in no. 1, it is necessary to first develop some material on Cartan subgroups and parabolic groups. Let $\mathfrak{g}$ be a real semisimple Lie algebra, $\mathfrak{g}_c$ its complexification. A subalgebra $\mathfrak{h} \subseteq \mathfrak{g}$ is called a *Cartan subalgebra* if its complexification $\mathfrak{h}_c$ is a Cartan subalgebra of $\mathfrak{g}_c$. Let $\mathfrak{g} = \mathfrak{k} + \mathfrak{p}$ be a Cartan decomposition of $\mathfrak{g}$ and θ the corresponding Cartan involution.

LEMMA 7. (Harish-Chandra [4]) *Let* $\mathfrak{h}$ *be a Cartan subalgebra of* $\mathfrak{g}$ *. Then it is possible to conjugate* $\mathfrak{h}$ *by an element of* Ad $\mathfrak{g}$ *so that the resulting Cartan subalgebra is* θ*-invariant.*

We assume henceforth that $\mathfrak{h}$ is θ-invariant. This means in particular that

$$\mathfrak{h} = (\mathfrak{h} \cap \mathfrak{k}) + (\mathfrak{h} \cap \mathfrak{p}).$$

The reader should be aware that not all Cartan subalgebras of $\mathfrak{g}$ are Ad $\mathfrak{g}$ -conjugate in general. However, there are only finitely many conjugacy classes. Moreover all Cartan subalgebras have the same dimension, namely rank $\mathfrak{g}$. Note that for a θ-invariant subalgebra, we may assume $\mathfrak{h} \cap \mathfrak{p} \subseteq \mathfrak{a}$. There is only one class wherein $\mathfrak{h} \cap \mathfrak{p} = \mathfrak{a}$, the *maximally split* Cartan subalgebra. In such a case, $\mathfrak{h} \cap \mathfrak{k}$ is a Cartan subalgebra of $\mathfrak{m}$. The opposite extreme occurs when $\mathfrak{h} = \mathfrak{h} \cap \mathfrak{k}$ -- that is, the situation of the discrete series. In general, there is only one class of $\mathfrak{h}$ such that dim $(\mathfrak{h} \cap \mathfrak{k})$ is maximal. But there may exist non-conjugate $\mathfrak{h}_1$, $\mathfrak{h}_2$ such that dim $\mathfrak{h}_1 \cap \mathfrak{k} =$ dim $\mathfrak{h} \cap \mathfrak{k}$ and dim $\mathfrak{h}_1 \cap \mathfrak{p} =$ dim $\mathfrak{h}_2 \cap \mathfrak{p}$. Finally if there is an $\mathfrak{h}$ such that $\mathfrak{h} \cap \mathfrak{p} = \mathfrak{h}$, then $\mathfrak{g}$ is said to be *split* (and is then a normal real form of $\mathfrak{g}_c$).

EXAMPLES. (1) $sl(n,\mathbb{R})$ is split since we may take $\mathfrak{h}$ = the subalgebra of diagonal matrices.

(2) $so(n,1) = LA(SO(n,1))$, $n > 2$, is not split. The maximally split Cartan subalgebra looks like $\mathfrak{h} = \mathfrak{b} + \mathfrak{a}$ where $\mathfrak{b}$ is a maximal torus of $\mathfrak{m} = so(n-1)$ and $\mathfrak{a} \subseteq \mathfrak{p}$ has dimension 1.

DEFINITION. By a *Cartan subgroup* of a connected semisimple Lie group G we mean the centralizer of a Cartan subalgebra $\mathfrak{h}$ of $\mathfrak{g} = LA(G)$.

LEMMA 8. (Harish-Chandra [5]) *Let* H *be a Cartan subgroup of* G, H^0 *the neutral component*, $Z = \text{Cent } G$. *Then* H^0Z *is an open subgroup of* H *lying in the center of* H, *and* H/H^0Z *is finite. Also* H/Z *is abelian and if* G *has a faithful matrix representation,* H *is abelian. In general* $H \neq H^0$.

EXERCISE. Compute H when $G = SL(n,\mathbb{R})$ and $\mathfrak{h}$ = the diagonal matrices in $\mathfrak{g}$.

Next we turn to the topic of parabolic groups. A *parabolic group* $P \subseteq G$ is by definition a closed subgroup of G such that (1) if $\mathfrak{p}' = LA(P)$, then $P = N(\mathfrak{p}')$ and (2) $\mathfrak{p}'_c$ contains a maximal solvable subalgebra of $\mathfrak{g}_c$ (i.e. somthing of the form $\mathfrak{h}_c + \sum_{\alpha\in\Delta} \mathfrak{g}_c^{\alpha}$). Let N be the maximal normal subgroup of P such that Ad(n) is unipotent for all $n \in N$. Let $\Xi = P \cap \tilde{\theta}P$. Next set A = the maximal connected split (i.e., Ad(a) diagonalizable over $\mathbb{R}$) abelian subgroup contained in Cent Ξ. Then Ξ is the centralizer of A in G. A is called the *split component* of P. Next let $X(\Xi) = \{\chi: \Xi \to R^*$, χ a continuous homomorphism$\}$. Set $M = \bigcap_{\chi\in X(\Xi)} \ker|\chi|$. Then M is reductive (that is, LA(M) is reductive) and not connected in general. Moreover $\Xi = MA$ is a direct product and $P = MAN$ is a semidirect product. The latter is called the *Langlands decomposition* of P and agrees with the MAN decomposition of the minimal parabolic B. (Note: the rationale behind the way that M was obtained will become clear only after we study algebraic groups -- see Chapter V, Theorems A3, 4 and 6.)

DEFINITION. P is called *cuspidal* if there is a θ-stable Cartan subalgebra $\mathfrak{h}$ such that $\mathfrak{h} \cap \mathfrak{p} = \mathfrak{a} = LA(A)$.

If P is cuspidal then $H = Z(\mathfrak{h})$ is a Cartan subgroup of G which we say is *compatible* with P. Set $B = H \cap K$. Then $H = BA$ is a direct product, although neither H nor B need necessarily be connected or abelian.

EXERCISE. (Lipsman [3]) Show that B is a compact Cartan subgroup of M, that is, show that $B = Z(\mathfrak{b}) \cap M$, where $\mathfrak{b} = \mathfrak{h} \cap \mathfrak{k}$ is a Cartan subalgebra of $\mathfrak{m} = LA(M)$.

Set $\mathfrak{z} = LA(\Xi) = \mathfrak{m} + \mathfrak{a}$. $\mathfrak{z}$ is reductive and θ-stable $\Longrightarrow$ $\mathfrak{z} = (\mathfrak{z} \cap \mathfrak{k}) + (\mathfrak{z} \cap \mathfrak{p})$. Put $M_{\mathfrak{p}} = \exp(\mathfrak{m} \cap \mathfrak{p})$. Then $G = KP$ and the map

$$(k,m,a,n) \to kman, \qquad K \times M_{\mathfrak{p}} \times A \times N \to G$$

is an analytic diffeomorphism.

Next let $\mathfrak{g} \subseteq \mathfrak{g}_c$ and identify $Ad(\mathfrak{g})$ with the analytic subgroup of $Ad(\mathfrak{g}_c)$ corresponding to $ad(\mathfrak{g}) \subseteq ad(\mathfrak{g}_c)$. Set

$$\Gamma = Ad_G^{-1}(Ad_G(K) \cap \exp i\mathfrak{a}).$$

THEOREM 9. (Lipsman [3]) (i) Γ is a finite subgroup of B that commutes with M^0.

(ii) Γ is normal in M.

(iii) $H = H^0\Gamma$.

Proof. (i) $\ker Ad_G = Z_G$ is a finite group and $Ad_G(K) \cap \exp i\mathfrak{a}$ finite $\Longrightarrow$ Γ is also finite. Γ is contained in M and clearly centralizes $\mathfrak{b} \Longrightarrow \Gamma \subseteq B$. Since $[\mathfrak{a}, \mathfrak{m}] = 0$, Γ and M^0

must commute.

(ii) Exercise.

(iii) We give the proof for the special case $Z_G = \{e\}$. It suffices to prove $B = B^0\Gamma$. Let H_c be the Cartan subgroup of $\mathrm{Ad}\,\mathfrak{g}_c$ with Lie algebra $\mathfrak{h}_c$. Set $\mathfrak{u} = \mathfrak{k} + i\mathfrak{p}$. Then $U = \exp\mathfrak{u}$ is a compact group. In fact, it is the maximal compact subgroup of $G_c = \mathrm{Ad}\,\mathfrak{g}_c$. But $U \cap H_c = U \cap Z(\mathfrak{h}_c)$ and $\mathfrak{h}_c$ is invariant under the Cartan involution of $\mathfrak{g}_c$ determined by $\mathfrak{u}$. By Harish-Chandra [6, Lemma 27] it follows that $U \cap H_c$ is connected and $\exp(\mathfrak{u} \cap \mathfrak{h}_c) = U \cap H_c$. But $\mathfrak{u} \cap \mathfrak{h}_c = (\mathfrak{h} \cap \mathfrak{k}) + i(\mathfrak{h} \cap \mathfrak{p}) = \mathfrak{b} + i\mathfrak{a}$. Let $b \in B \subseteq U \cap H_c$. Then $b = b_1 b_2$, where $b_1 \in B^0$ and $b_2 = b_1^{-1}b \in \exp i\mathfrak{a} \cap K = \Gamma$.

REMARK. When P is minimal, M is compact and $M = M^0\Gamma$. Neither is true in general for arbitrary cuspidal parabolics. However the entire theory of the discrete series can be carried over to the group M (Lipsman [3]). The procedure consists of extending Harish-Chandra's theory to connected reductive Lie groups with compact center, then to the direct product $M^0\Gamma$, and then to use the Mackey extension procedure (which is discussed in Chapter III) to obtain the discrete series for the group M. For more detail the reader is referred to Lipsman [3].

EXAMPLE. $G = SL(3,\mathbb{R})$, $P = \left\{\begin{pmatrix} a & b & c \\ d & e & f \\ 0 & 0 & g\end{pmatrix} : (ae-bd)g = 1\right\}$,

$$N = \left\{\begin{pmatrix} 1 & 0 & e \\ 0 & 1 & f \\ 0 & 0 & 1\end{pmatrix} : e,f \in \mathbb{R}\right\}, \quad \Xi = \left\{\begin{pmatrix} a & b & 0 \\ d & e & 0 \\ 0 & 0 & g\end{pmatrix} : (ae-bd)g = 1\right\},$$

$$A = \left\{\begin{pmatrix} a & 0 & 0 \\ 0 & a & 0 \\ 0 & 0 & a^{-2}\end{pmatrix} : a > 0\right\}, \quad M = \left\{\begin{pmatrix} a & b & 0 \\ d & e & 0 \\ 0 & 0 & g\end{pmatrix} : ae-bd = \pm 1\right\},$$

$$\mathfrak{h} = \left\{\begin{pmatrix} a & b & 0 \\ -b & a & 0 \\ 0 & 0 & -2a\end{pmatrix} : a,b \in \mathbb{R}\right\}, \quad B = B^0 = \left\{\begin{pmatrix} \cos\theta & \sin\theta & 0 \\ -\sin\theta & \cos\theta & 0 \\ 0 & 0 & 1\end{pmatrix} : \theta \in \mathbb{R}\right\},$$

$\Gamma = \{e\}$. In this case $M \neq M^0\Gamma$.

For our last definition, set $\Xi' = N(A)$. Then Ξ'/Ξ is a finite group denoted by W_A. W_A acts on $\hat{A}$ and $\hat{M}$ exactly as in the minimal case. The space $P\backslash G/P$ is of course finite (for more information on it see Chapter V, Theorem A 11). The *principal series* corresponding to P (or P-series) is obtained as follows. Let $\sigma \in \hat{M}_d$, $\tau \in \hat{A}$. Form the representation $\sigma \times \tau$ of P given by $(\sigma \times \tau)(man) = \sigma(m)\tau(a)$. Then set $\pi(\sigma,\tau) = \mathrm{Ind}_P^G\ \sigma \times \tau$. These representations can be realized on function spaces over both $K/K \cap M$ and $\tilde{\theta}N$ as we did in the minimal case. The exact details are left to the reader as an exercise (in particular to compute the q-function q_P that gives the quasi-invariant measure on G/P).

THEOREM 10. (i) $\mathscr{I}(\pi(\sigma_1,\tau_1),\pi(\sigma_2,\tau_2)) \leqq \#s \in W_A\colon s(\sigma\times\tau) \cong \sigma\times\tau$.
(ii) $\pi(s\sigma,s\tau) \cong \pi(\sigma,\tau)$, $s \in W_A$.

Result (ii) appears in Harish-Chandra [11]. A weaker version of (i) also appears there. That result (i) as stated here actually holds was told to me by Harish-Chandra. See Chapter V, Theorem C 5 for a corresponding version in the p-adic case. So once again almost all these principal series representations are irreducible. Much less is known about the reducible ones than in the minimal case.

EXERCISE. For $G = SL(n,\mathbb{R})$, compute which of the parabolic groups $P = \begin{pmatrix} \boxed{*} & & & * \\ & \boxed{*} & & \\ & & \boxed{*} & \\ 0 & & & \boxed{*} \end{pmatrix}$ are cuspidal, and write down realizations of the corresponding principal series representations.

4. <u>Complementary series</u>. There are other families of irreducible representations of semisimple Lie groups beside the discrete series and various principal series. We discuss two more cases, one in this no., one in the next.

The first is the complementary series discovered initially by Gelfand and Naimark. It can be described in a general framework as follows. Let G be a connected semisimple Lie group with finite center, and let P be a cuspidal parabolic subgroup with $P = MAN$ its Langlands decomposition. Take $\sigma \in \hat{M}_d$ and let τ be a non-unitary character of A, $\tau: A \to \mathbb{C}^*$. Begin by formally writing down one of the realizations of the "induced representation" $\pi(\sigma,\tau)$; for example, let

$$\mathcal{H}_{\pi(\sigma,\tau)} = \Big\{ f: G \to \mathcal{H}_\sigma, \quad f \text{ meas.}, \quad f(mang) = q_P^{\frac{1}{2}}(a)\tau(a)\sigma(m)f(g), \quad \int_{K/K\cap M} \|f(\bar{k})\|^2 d\bar{k} < \infty \Big\}$$

with the representation acting via

$$\pi(\sigma,\tau)(g)f(x) = f(xg), \quad g,x \in G, \quad f \in \mathcal{H}_{\pi(\sigma,\tau)}.$$

The result is a bounded (non-unitary) representation of G in Hilbert space. However it *may* be possible to find a new inner product $(\cdot,\cdot)_{\sigma,\tau}$ on $\mathcal{H}_{\pi(\sigma,\tau)}$, or on some dense subspace $\mathcal{D}$, with respect to which the operators $\pi(\sigma,\tau)$ become isometries. If that is possible, we refer to the resulting unitary representation on the completion of $\mathcal{D}$ (with respect to $(\cdot,\cdot)_{\sigma,\tau}$) as a *complementary series* representation.

EXAMPLE. Let $G = SL(2,\mathbb{R})$ and take $P = \Big\{ \begin{pmatrix} a & b \\ 0 & a^{-1} \end{pmatrix} : a \neq 0,\ b \in \mathbb{R} \Big\}$. If $g = \begin{pmatrix} a & b \\ c & d \end{pmatrix}$, then

$$\pi(\varepsilon,\tau)(g)f(x) = |bx+d|^{-1+\tau}\ \mathrm{sgn}(bx+d)^{\varepsilon}\ f\Big(\frac{ax+c}{bx+d}\Big), \qquad f \in L_2(\mathbb{R})$$

where $\tau \in \mathbb{C}$, $\varepsilon = 0,1$. This gives a principal series representation if $\tau \in i\mathbb{R}$. If on the other hand we let $\tau \in \mathbb{R}$, $0 < \tau < 1$, $\varepsilon = 0$ and define the inner product

$$(f,h)_\tau = \iint |x-y|^{-\tau} f(x)\bar{h}(y)\,dx\,dy\ ,$$

we obtain representations π_τ which constitute the complementary series for G.

THEOREM 11. (i) (Lipsman [2]) *In general, complementary series representations exist.*

(ii) (Kostant [1]) *Suppose* P *is minimal,* $\sigma = 1_M$, $\tau \in CS$ = *a certain critical strip around the imaginary axis (on which the principal series is defined), and there is* $s \in W$ *such that* $s\tau = \bar{\tau}^{-1}$. *Then the representation* $\pi_\tau = \pi(1_M,\tau)$ *exists as a complementary series representation.*

It is an understatement to say that the complementary series is not particularly well understood at this stage.

5. Degenerate series. Regarding representations induced from a parabolic subgroup $P = MAN$, there are two natural questions that one can ask: (1) What if we use irreducible representations of M other than square-integrable ones; and (2) What if P is not cuspidal? The answer to (1) is that one obtains series of representations which are the same as those induced as usual from "smaller" cuspidal parabolics (i.e., parabolics with larger split components). The answer to (2) is that one obtains the so-called *degenerate series*.

EXAMPLE. Let $G = SL(3,\mathbb{C})$, $P = \left\{\begin{pmatrix} \alpha & \beta & \varepsilon \\ \gamma & \delta & \phi \\ 0 & 0 & \eta \end{pmatrix} : \text{complex entries}\right\}$.

The Langland's decomposition is $P = MAN$ where

$$M = \left\{\begin{pmatrix} \alpha & \beta & 0 \\ \gamma & \delta & 0 \\ 0 & 0 & e^{i\phi} \end{pmatrix} : (\alpha\delta-\beta\gamma)e^{i\phi} = 1\right\}, \quad A = \left\{\begin{pmatrix} \alpha & 0 & 0 \\ 0 & \alpha & 0 \\ 0 & 0 & \alpha^{-2} \end{pmatrix} : \alpha > 0\right\},$$

$$N = \left\{\begin{pmatrix} 1 & 0 & \varepsilon \\ 0 & 1 & \phi \\ 0 & 0 & 1 \end{pmatrix} : \varepsilon,\phi \in \mathbb{C}\right\}.$$

Clearly $M \cong \mathbb{T}\cdot SL(2,\mathbb{C})$, a reductive group which does not have a compact Cartan subgroup. Hence P is not cuspidal. Consider $\pi(\sigma,\tau)$, $\sigma \in \hat{M}$, $\tau \in \hat{A}$. If $\sigma = \sigma_1 \times \sigma_2$, where $\sigma_1 \in \hat{\mathbb{T}}$, σ_2 a principal series representation of $SL(2,\mathbb{C})$, we get an ordinary principal series representation of $SL(3,\mathbb{C})$. However if we take $\sigma_2 = 1$, then we get degenerate series representations of $SL(3,\mathbb{C})$.

In general now if $P = MAN$ is a non-cuspidal parabolic, $\sigma \in \hat{M}$ is a character, and $\tau \in \hat{A}$, the representations $\pi(\sigma,\tau) = \mathrm{Ind}_P^G\, \sigma \times \tau$ are called the *degenerate series* (corresponding to P).

THEOREM 12 (folklore). *Almost all the degenerate series is irreducible.*

These representations have been studied (see Gelfand and Naimark [1] and Gross [1]), but not very extensively. We shall see in the next section one reason for this -- namely they do not occur in the decomposition of the regular representation of G. The same is true of the complementary series. Finally, we leave it to the reader to ponder the possibility of complementary degenerate series or degenerate complementary series or any other bizarre family of representations his fancy leads him to.

C. CHARACTERS AND THE PLANCHEREL FORMULA

In this section we investigate the character theory of the representations we have constructed and see how it is related to the so-called Plancherel measure. We first summarize the facts concerning the general Plancherel Theorem (see Dixmier [5]).

Let G be a locally compact group which is *type I*. For the precise definition of type I see Dixmier [5]. Perhaps a good way to think of it is that the following is true: any $\omega \in \mathrm{Rep}(G)$ determines a *unique* measure class $\{\mu_\omega\}$ and a *unique* multiplicity function $m_\omega(\pi)$ such that

$$\omega = \int_{\hat{G}}^{\oplus} m_\omega(\pi)\,\pi\,d\mu_\omega(\pi),$$

and conversely all unitary representations are obtained in this way.

Next suppose G is unimodular and ω is the left regular representation λ_G of G

$$\lambda_G(g)f(x) = f(g^{-1}x), \quad f \in L_2(G), \quad x,y \in G.$$

Then we have

THEOREM 1. *The multiplicity function* m_{λ_G} *is given by* $m_{\lambda_G}(\pi) = \dim \pi$ *and there is a unique choice of* $\mu_G \in \{\mu_{\lambda_G}\}$ *(depending only on the normalization of Haar measure on* G*) such that*

$$\lambda_G \cong \int_{\hat{G}}^{\oplus} (\dim \pi)\,\pi\,d\mu_G(\pi)$$

$$\int_G |f(g)|^2 dg = \int_{\hat{G}} \|\pi(f)\|_2^2 \, d\mu_G(\pi),$$

where $\|\cdot\|_2$ *denotes the Hilbert-Schmidt norm.*

Three of the most important problems in representation theory are: (1) for a given locally compact group G, determine if it is type I; (2) if so, describe $\hat{G}$ as completely and explicitly as possible; and (3) in the unimodular case compute μ_G.

When G is a semisimple Lie group, it is known that G is type I (see later discussion in this section). Also a great deal is known about $\hat{G}$ (see section B), in fact enough so that the Plancherel measure has been computed by Harish-Chandra. We proceed to give a description of the measure, and then say a few words about the method of obtaining it and the role characters play.

So as usual G denotes a connected semisimple Lie group with finite center. Suppose P_1, P_2 are two cuspidal parabolic subgroups of G, with split components A_1, A_2. We say P_1 and P_2 are *associate* if there is $x \in G$ such that $xA_1x^{-1} = A_2$. It follows that if $P = MAN$ is a cuspidal parabolic, then (up to conjugacy) any other associate parabolic P' looks like $P' = MAN'$.

LEMMA 2. (i) (Lipsman [3]) *Let* P_1, P_2 *be two cuspidal parabolics;* H_1, H_2 *corresponding compatible Cartan subgroups. Then* P_1 *and* P_2 *are associate if and only if* H_1 *and* H_2 *are conjugate.*

(ii) (Harish-Chandra [4]) *There are only finitely many conjugacy classes of Cartan subgroups.*

(iii) (Harish-Chandra [9]) *If* $P_1 = MAN_1$, $P_2 = MAN_2$ *are associate cuspidal parabolics,* $\sigma \in \hat{M}_d$, $\tau \in \hat{A}$, *the representations* $\pi^1(\sigma,\tau) = \mathrm{Ind}_{P_1}^G\ \sigma \times \tau$ *and* $\pi^2(\sigma,\tau) = \mathrm{Ind}_{P_2}^G\ \sigma \times \tau$ *are equivalent.*

(iv) (Lipsman [3]) *If* P_1, P_2 *are two non-associate cuspidal parabolics, then any representation in the* P_1*-series is not equivalent to (in fact is disjoint from) any representation in the* P_2*-series.*

Let $P_1, \ldots, P_r$ be a complete set of non-associate proper cuspidal parabolics, $P_i = M_iA_iN_i$. Set $P_0 = G$ if rank K = rank G. Let $H_0, H_1, \ldots, H_r$ be a corresponding set of Cartan subgroups (if $P_0 = G$, then H_0 is a compact Cartan subgroup).

THEOREM 3. (Harish-Chandra [9]) *There are continuous functions of polynomial growth*

$$C_0(\lambda), \quad \lambda \in \mathcal{L}' \cap \mathcal{F}$$

$$C_i(\sigma_i,\tau_i), \quad \sigma_i \in \hat{M}_{i_d}, \quad \tau_i \in \hat{A}_i, \quad 1 \leqq i \leqq r,$$

which are invariant under the appropriate Weyl group (W_B *for* $i = 0$, W_{A_i} *for* $1 \leqq i \leqq r$), *such that* $d\mu_G = C_0(\lambda)d\lambda + \sum_{i=1}^{r} C_i(\sigma_i,\tau_i)d\sigma_i \times d\tau_i$,

$d\sigma_i$ = *the counting measure on* $\hat{M}_{i_d}$, $d\tau_i$ = *Lebesgue measure on* $\hat{A}_i$, $d\lambda$ = *counting measure on* $\mathcal{L}' \cap \mathcal{F}$.

Recall that $\hat{M}_{i_d}$ is parameterized by part of a certain lattice in a vector space (at least for M_i connected and semisimple -- it is essentially true for general M_i as well, see Theorem 5). Thus it makes sense to speak of polynomial-growth functions on $\hat{M}_i$. The reader is referred to Harish-Chandra [11] for more information on the C functions, in particular for their expression as a product of quotients of Gamma functions.

Put in as elementary fashion as possible, Harish-Chandra's method of proof for Theorem 3 is the following: Setting $h = f^* * f$ in the Plancherel formula it becomes

$$h(e) = \int_{\hat{G}} \mathrm{Tr}\ \pi(h) d\mu_G(\pi).$$

We shall refer to this equation as "the inversion formula". If one can prove the inversion formula for a sufficiently large collection of h's, then it follows that μ_G is the Plancherel measure (see Warner [1]). The collection Harish-Chandra uses is $C_0^\infty(G)$.

DEFINITIONS. (1) A Lie group G is called *traceable* if for every $f \in C_0^\infty(G)$ and every $\pi \in \hat{G}$, the operator $\pi(f)$ is trace class and $f \to \mathrm{Tr}\ \pi(f)$ is a distribution on G.

(2) G is called *CCR* if for every $f \in L_1(G)$ and every $\pi \in \hat{G}$, the operator $\pi(f)$ is compact.

We have CCR $\Longrightarrow$ type I (see Dixmier [5]), but not conversely. Also traceable $\Longrightarrow$ CCR, but it is an open problem as to whether the converse is true or not.

The fact that a semisimple group is traceable is obtained by the following method. A compact group K in a locally compact group G is called *large* if $\forall \pi \in \hat{G}$, $\forall \omega \in \hat{K}$, the number of times the representation $\pi|_K$ contains ω is finite.

THEOREM 4. (Harish-Chandra [1]) *If* G *is a connected semisimple Lie group with finite center and* K *is a maximal compact subgroup, then* K *is large in* G.

Using this theorem and some relatively easy growth estimates (see Harish-Chandra [1]), we find the

COROLLARY. *Connected semisimple Lie groups having finite center are traceable.*

The distributions $\Theta_\pi: f \to \mathrm{Tr}\ \pi(f)$ are called the *characters* of G. In order to prove Theorem 3, one works with the characters rather than the representations themselves. Essentially, one computes the characters of the discrete series (a formidable task -- see Harish-Chandra [7,8]), and then the characters of the various principal series (considerably less formidable -- see e.g. Lipsman [3]), and then (by some very difficult analysis -- see Harish-Chandra [11] for an outline) one derives the inversion formula for G. We shall not describe the analysis that proceeds from the characters to the Plancherel formula. However, we shall go into more detail on the characters themselves. First we explain how the inversion formula can be thought of as an eigenvalue expansion, and then we give explicit character formulas.

If G is a Lie group, $\mathfrak{g} = LA(G)$, and $\pi \in Rep(G)$, then $\xi \in \mathcal{H}_\pi$ is called C^∞ if $g \to \pi(g)\xi$, $G \to \mathcal{H}_\pi$, is infinitely differentiable. The collection $C^\infty(\pi)$ of vectors thus determined is a dense subspace of $\mathcal{H}_\pi$ which is a $\mathfrak{g}$-module via

$$\pi(X)\xi = \frac{d}{dt}\pi(\exp tX)\xi|_{t=0}.$$

This gives a representation of $\mathfrak{g}$ in $\mathcal{H}_\pi$ as unbounded skew hermitian operators. It can be extended to the *universal enveloping algebra* $\mathfrak{U}(\mathfrak{g})$ in a canonical way. By $\mathfrak{U}(\mathfrak{g})$ we mean the associative algebra generated by $\mathfrak{g}$, that is (in terms of a basis X_i of $\mathfrak{g}$) the algebra $\mathbb{C}[X_1,\ldots,X_n]/R_0$ where R_0 is the two-sided ideal generated by the polynomials $X_iX_j - X_jX_i - [X_i,X_j]$. It is well-known that $\mathfrak{U}(\mathfrak{g})$ is isomorphic to the algebra of left-invariant differential operators on G. It is also possible to identify $\mathfrak{U}(\mathfrak{g})$ (as a linear space) with polynomial functions on $\mathfrak{g}^* = \mathrm{Hom}_R(\mathfrak{g},\mathbb{R})$. Then if we set $\mathfrak{Z}(\mathfrak{g}) = \mathrm{Cent}\,\mathfrak{U}(\mathfrak{g})$, we have: $p \in \mathfrak{U}(\mathfrak{g})$ is in $\mathfrak{Z}(\mathfrak{g})$ $\Longleftrightarrow$ p is invariant under the co-adjoint representation. Now π extends to $\mathfrak{U}(\mathfrak{g})$ naturally and satisfies

$$\pi(pf) = \pi(p)\pi(f), \quad p \in \mathfrak{U}(\mathfrak{g}), \quad f \in C_0^\infty(G).$$

Of course for $z \in \mathfrak{Z}(\mathfrak{g})$, the operators $\pi(z)$ commute with the operators of the representation. In particular if π is irreducible, then $\pi(z) = \chi_\pi(z)I$, $z \in \mathfrak{Z}(\mathfrak{g})$. The homomorphism $\chi_\pi : \mathfrak{Z}(\mathfrak{g}) \to \mathbb{C}$ is called the *infinitesimal character* of π. As a rule, representations $\pi \in \hat{G}$ are not uniquely determined by their infinitesimal characters.

Now assume the group G is traceable. Then for $\pi \in \hat{G}$, $f \in C_0^\infty(G)$, $z \in \mathfrak{Z}(\mathfrak{g})$, we can compute

$$(z\Theta_\pi)(f) = \Theta_\pi(z^* f) = \mathrm{Tr}\ \pi(z^* f)$$

$$= \mathrm{Tr}\ \pi(z)^*\pi(f) = \overline{\chi_\pi(z)}\ \Theta_\pi(f).$$

That is, the characters of irreducible representations are eigendistributions of $\mathfrak{Z}(\mathfrak{g})$, with infinitesimal character as eigenvalue. Thus the inversion formula can be rewritten

$$\delta = \int_{\hat{G}} \Theta_\pi d\mu_G(\pi)$$

and thought of as an eigenvalue expansion of the Dirac distribution δ, $\delta(f) = f(e)$, $f \in C_0^\infty(G)$.

Now resume the assumption G semisimple. Let $n = \dim \mathfrak{g}$, $\ell = \dim \mathfrak{h}$, where $\mathfrak{h} \subseteq \mathfrak{g}$ is a Cartan subalgebra. Fix an indeterminate t and consider $\det(t + 1 - \mathrm{Ad}_G(x)) = D_0(x) + \ldots + D_n(x)t^n$, $x \in G$. The first non-zero coefficient will be $D_\ell(x)$; set $D(x) = D_\ell(x)$. Then by the *regular elements* G' of G we mean the dense open submanifold $G' = \{x \in G: D(x) \neq 0\}$.

EXERCISE. Check that $D(xz) = D(x)$, $z \in Z_G$, and $D(gxg^{-1}) = D(x)$, $g,x \in G$.

For any subset S, we set $S' = S \cap G'$. Then if H is a Cartan subgroup of G, consider the map

$$\phi_H: Z_H\backslash G \times H' \to G', \quad \phi_H(\bar{g},h) = g^{-1}hg, \quad \bar{g} = Z_H g.$$

The image, denoted G_H', is an open submanifold of G, ϕ_H is proper and G' is a disjoint union of the G_{H_i}' as H_i varies over a complete set of non-conjugate Cartan subgroups. In addition if $W_H = N(H)/Z_H$, W_H acts effectively on $Z_H\backslash G \times H'$ and G_H' is diffeomorphic to $(Z_H\backslash G \times H')/W_H$.

Next let $\lambda: \mathfrak{h} \to \mathbb{C}$ be a linear form. Then there exists at most one homomorphism $\xi_\lambda: H \to \mathbb{C}^*$ such that $\xi_\lambda(\exp Y) = e^{\lambda(Y)}$, $Y \in \mathfrak{h}$. If α is a root of $(\mathfrak{g}_c, \mathfrak{h}_c)$ restricted to $\mathfrak{h}$, ξ_α always exists. Let Q denote a choice of positive roots for $(\mathfrak{g}_c, \mathfrak{h}_c)$ and set $\rho = \frac{1}{2}\sum_{\alpha \in Q} \alpha$. G is called *acceptable* if ξ_ρ exists. By passing to a finite covering we may always assume G is acceptable. Set

$$\Delta_H(h) = \xi_\rho(h) \prod_{\alpha \in Q} (1 - \xi_\alpha(h^{-1})), \quad h \in H$$

$$\Delta_H^+(h) = \xi_\rho(h_+) \prod_{\alpha \in Q_+} (1 - \xi_\alpha(h^{-1})),$$

where $h = h_- h_+$ is the unique decomposition of h into BA, $h_+ \in A$ and $Q_+ = \{\alpha \in Q: \alpha|_{\mathfrak{a}} \neq 0\}$.

THEOREM 5. (i) (Harish-Chandra [6]) *For* $\pi \in \hat{G}$, Θ_π *is actually given by a locally integrable function.*

(ii) (Harish-Chandra [7,8] *Let* G *have a compact Cartan subgroup* B. *Then on* B *the character of the representation* π_λ, $\lambda \in \mathcal{L}' \cap \mathcal{F}$, *of the discrete series* (Theorem B6) *has the form*

$$\Theta_\lambda(b) = \frac{c}{\Delta_B(b)} \sum_{W_B} \varepsilon(s) e^{s\lambda(Y)}, \qquad \exp Y = b \in B',$$

$c = (-1)^q \operatorname{sgn} \omega(H_\lambda)$, $q = \frac{1}{2} \dim G/K$ *and* $\omega^s = \varepsilon(s)\omega$, $s \in W_B$.

(iii) (Lipsman [3]) *Result* (ii) *can be extended to groups* M *that occur in cuspidal parabolics* $P = MAN$.

(iv) (Lipsman [3]) *For* $\pi(\sigma,\tau)$ *in the* P-*series, the character has the following form on a compatible Cartan subgroup* H

$$\Theta_{\pi(\sigma,\tau)}(h) = \frac{c}{|\Delta_+(h)|} \sum_{W_H} \Theta_\sigma(sh)\tau(sh), \qquad h \in H'.$$

REMARK. Let $P = MAN$ be a cuspidal parabolic. In Theorem B9 we have seen that the group $W_A = N(A)/Z(A)$ determines equivalence among the P-series representations. However part (iv) of Theorem 5 says that $W_H = N(H)/Z_H$ does likewise. We can reconcile this apparent discrepancy with the following observation:

From the facts that $N(A) = (N(A) \cap K)A$ and that any two compact Cartan subgroups of M are conjugate in M (Lipsman [3]), it follows easily that $N(A) = N(H)Z(A)$. But then

$$\begin{aligned} W_A &= N(A)/Z(A) = N(H)Z(A)/Z(A) \\ &= N(H)/N(H) \cap Z(A) = (N(H)/Z_H)/\{(N(H) \cap Z(A)/Z_H\} \\ &= W_H/W_0 \end{aligned}$$

where $W_0 = (N(H) \cap Z(A))/Z_H$. Obviously the group W_0 acts trivially on $\hat{A}$. Although W_0 acts non-trivially on M -- in fact, by inner automorphism -- it therefore acts trivially on $\hat{M}$. (Note: This argument generalizes the corresponding argument for minimal parabolics found in Lipsman [1].)

We make a final comment. For various reasons (to some extent similar to those that arise in abelian Fourier analysis), the space $C_0^\infty(G)$ is too small to support the analysis necessary to obtain μ_G. One has to use a larger space of functions. For abelian analysis that role is played by the rapidly decreasing C^∞ functions. For semi-simple groups Harish-Chandra employs his *Schwartz space* $\mathcal{S}(G)$ (see Harish-Chandra [8] for the definition).

THEOREM 6. (Harish-Chandra [8,9]) *Let* $\pi \in \hat{G}$ *be either a principal or discrete series representation. Then* Θ_π *is a* tempered distribution, *that is, it extends to a continuous linear functional on the Schwartz space* $\mathcal{S}(G)$.

REMARK. It has been conjectured that a converse of Theorem 6 should hold, that is, if Θ_π, $\pi \in \hat{G}$, is tempered, then π should be in the support of the regular representation. That is apparently still an open question.

We close Chapter I with a standard example and some current areas of study.

EXAMPLE. Let $G = SL(2,\mathbb{R})$. G has two conjugacy classes of Cartan subgroups and we may take as representatives

$$H = \left\{\begin{pmatrix} h & 0 \\ 0 & h^{-1}\end{pmatrix} : h \in \mathbb{R}^*\right\}, \quad B = \left\{\begin{pmatrix} \cos\phi & \sin\phi \\ -\sin\phi & \cos\phi\end{pmatrix} : \phi \in \mathbb{R}\right\}. \text{ Here}$$

$B = K$ is a maximal compact subgroup (a rare situation admittedly). The group H corresponds to the minimal parabolic

$P = \left\{\begin{pmatrix} a & b \\ 0 & a^{-1}\end{pmatrix} : a \in \mathbb{R}^*,\ b \in \mathbb{R}\right\}$. The representations of the principal series are $\pi(\sigma,\tau) = \mathrm{Ind}_P^G\, \sigma \times \tau$, $\sigma \in \hat{M}$, $M = \begin{pmatrix} \pm 1 & 0 \\ 0 & \pm 1\end{pmatrix}$, $\tau \in \hat{A}$, $A = \left\{\begin{pmatrix} a & 0 \\ 0 & a^{-1}\end{pmatrix} : a > 0\right\}$. The characters of these representations are supported by the set G_H' and on the set H' are given by the formulas

$$\Theta_{\sigma,\tau}(h) = \frac{|a|^{i\rho} + |a|^{-i\rho}}{|a-a^{-1}|}\,\mathrm{sgn}^{\varepsilon}(a), \qquad h = \begin{pmatrix} a & 0 \\ 0 & a^{-1}\end{pmatrix} \in H'$$

if $\tau\begin{pmatrix} a & 0 \\ 0 & a^{-1}\end{pmatrix} = a^{i\rho}$, $\sigma\begin{pmatrix} -1 & 0 \\ 0 & -1\end{pmatrix} = \varepsilon = \pm 1$. These formulas reflect the fact that $\pi(\sigma,\tau) \cong \pi(\sigma,\tau^{-1})$, but do not reveal the (true) fact that $\pi(\sigma,\tau)$ is irreducible except when $\sigma \neq 1$, $\tau = 1$.

The discrete series representations of G (corresponding to the compact Cartan B) were written down in section B2. Recall they are parameterized by $\pi_n^{\pm}$, $n = 1, \frac{3}{2}, 2, \ldots$. The corresponding characters are supported by both G_H' and G_B'. The formulas are

$$\Theta_n^{\pm}(h) = \frac{a^{1-2n}}{a-a^{-1}}, \quad h = \begin{pmatrix} c & 0 \\ 0 & c^{-1}\end{pmatrix}, \quad a = c \text{ or } c^{-1} \text{ according as } |c| \text{ or } |c^{-1}| \text{ is larger}$$

$$\Theta_n^{\pm}(b) = \frac{e^{\mp i\phi(-1+2n)}}{\pm(e^{i\phi} - e^{-i\phi})}, \quad b = \begin{pmatrix} \cos\phi & \sin\phi \\ -\sin\phi & \cos\phi\end{pmatrix} \in B'.$$

Finally the Plancherel formula is well-known in this situation and is as follows: For a suitable normalization dg of Haar measure on G

$$\int_G |f(g)|^2 dg = \int_0^\infty \|\pi(1,\rho)(f)\|_2^2\, \rho \tanh \pi\rho \, d\rho + \int_0^\infty \|\pi(-1,\rho)(f)\|_2^2\, \rho \coth \pi\rho\, d\rho$$

$$+ \sum_{\substack{n \geq 1 \\ n \in \frac{1}{2}\mathbb{Z}}} (n-\tfrac{1}{2})(\|\pi_n^+(f)\|_2^2 + \|\pi_n^-(f)\|_2^2), \qquad f \in L_1(G) \cap L_2(G).$$

CURRENT TOPICS. (1) There is still no complete list of elements of $\hat{G}$ for other than a few cases like $SL(2,\mathbb{C})$, $SO_e(n,1)$.

(2) The representations in the discrete series have not been constructed in complete generality. An idea of Langlands, pursued by Schmid [1] and others, is to look for the representations in certain co-homology spaces. This has been fruitful, but has not yet yielded a complete solution.

(3) The computation of the Plancherel measure may be thought of as an explication of the L_2-analysis of G. It is only very recently (see Trombi and Varadarajan [1]) that people have begun seriously contemplating L_p-analysis, $p \neq 2$.

(4) Let $\Gamma \subseteqq G$ be a discrete subgroup such that G/Γ is compact or of finite volume. Compute the decomposition of the regular representation of G on $L_2(G/\Gamma)$! This is a large and important area of research, and is related to questions in automorphic forms, number theory, and the geometry of Lie groups.

(5) Consider results analogous to those we have presented for semisimple matrix groups over other locally compact fields -- see Chapter V, section B).

(6) Consider generalizing the results to semisimple adele groups.

CHAPTER II. RESULTS ON INDUCED REPRESENTATIONS

A. ANCIENT RESULTS OF MACKEY

While developing his idea of induced representations for infinite groups Mackey naturally tried to generalize many of the interesting results about finite groups. The outcome was his papers [3,4], many of whose theorems we describe here. We will give few complete proofs, but we will try to give a wide variety of interesting examples and applications. In particular we shall make elaborate use of the results of Chapter I.

1. Subgroup theorem. To begin we need the notions of measurable equivalence relation and regularly related subgroups. Let (X,μ) be a locally compact Hausdorff space, μ a finite regular Borel measure Let $\mathcal{R}$ be an equivalence relation on X, $Y = X/\mathcal{R}$ the set of equivalence classes, and $r\colon X \to Y$ the canonical projection.

DEFINITION. Say $\mathcal{R}$ is *measurable* if there is a countable collection $E_1, E_2, \ldots$ of subsets of Y such that $r^{-1}(E_i)$ is μ-measurable and every $y \in Y$ has the property that $\{y\} = \bigcap\{E_i\colon y \in E_i\}$.

Now let G be a locally compact group. We may consider Haar measure "to be finite" by taking an equivalent finite measure in its class. The concept of measurability for subsets of G is unaltered.

DEFINITION. Two closed subgroups G_1, G_2 of G are called *regularly related* if there exists a sequence $E_0, E_1, E_2, \ldots$ of measurable sets such that each E_i is a union of $G_1{:}G_2$ double cosets, E_0 has measure zero, and every double coset outside E_0 is the intersection of those E_j that contain it.

EXERCISE. Show that G_1 and G_2 are regularly related $\Longleftrightarrow$ the double cosets outside of a set of measure zero form the equivalence classes of a measurable equivalence relation.

It follows from Auslander and Moore [1, Prop. 2.12, p. 10] that if there is a Borel cross-section for the $G_1 : G_2$ double cosets, then G_1 and G_2 are regularly related.

DEFINITION. An *admissible measure* ν on $G_1 \backslash G/G_2$ is the following. Let μ be a finite Haar measure on G. We give $G_1 \backslash G/G_2$ the quotient Borel structure and set $\nu(E) = \mu(p^{-1}(E))$, where $p: G \to G_1 \backslash G/G_2$ is the canonical projection.

EXERCISE. Check that any two admissible measures are equivalent.

SPECIAL CASE. If there is a subset of G whose complement has measure zero, and which is the countable union of double cosets, then G_1 and G_2 are called *discretely related*. An admissible measure in such a case will be discrete.

EXAMPLES. (1) G connected semisimple Lie group, $G_1 = G_2 = P$ a parabolic group. Then $P\backslash G/P$ is finite, and so P is discretely related to itself.

(2) Suppose $G = HK$ where H and K are closed (not necessarily normal) subgroups of G. Then $H\backslash G/K$ has one point, and so H and K are discretely related. In particular the maximal compact subgroup K of a connected semisimple Lie group with finite center is discretely related to any parabolic subgroup P of G.

(3) G connected semisimple Lie group, $P = MAN$ a cuspidal parabolic, $V = \tilde{\theta}N$. Since PV has complement of measure zero in G, the groups P and V are discretely related.

(4) (Martin [1]) If G is an $\mathbb{R}$-rank one connected semisimple

Lie group with finite center, $P = MAN$ a minimal parabolic, then MA and MAN are regularly related (see section B).

THEOREM 1. (Subgroup Theorem - Mackey [3]). *Let* G_1 *and* G_2 *be regularly related in* G. *Let* $\pi \in \mathrm{Rep}(G_1)$. *For each* $x \in G$ *consider* $G_x = G_0 \cap (x^{-1}G_1x)$ *and set*

$$V_x = \mathrm{Ind}_{G_x}^{G}(\eta \to \pi(x\eta x^{-1})).$$

Then V_x *is determined to within equivalence by the double coset* $\bar{x}$ *to which* x *belongs. If* ν *is any admissible measure on* $G_1\backslash G/G_2$, *then*

$$\mathrm{Ind}_{G_1}^{G}\pi|_{G_2} \cong \int_{G_1\backslash G/G_2}^{\oplus} V_{\bar{x}}\,d\nu(\bar{x}).$$

Proof. (Sketch) Consider the space of the induced representation

$$\mathcal{H}(\pi) = \{f\colon G \to \mathcal{H}_\pi,\ \ f \text{ meas.},\ \ f(g_1g) = \pi(g_1)f(g),\ \ \int_{G/G_1} \|f(\bar{g})\|^2 d\bar{g} < \infty\}.$$

We denote by D a double coset and for $f \in \mathcal{H}(\pi)$ we write f_D for the restriction of f to D. One then shows essentially that the map

$$f \to \{f_D\}_{D\in G_1\backslash G/G_2}$$

is a unitary equivalence for the left and right sides of the final equation in the theorem.

EXAMPLES. (1) Let G be a connected semisimple Lie group with finite center, $G_1 = G_2 = P$ a minimal parabolic. Let $P = MAN$ and take $\sigma \in \hat{M}$, $\tau \in \hat{A}$, $\pi(\sigma,\tau) = \mathrm{Ind}_P^G\, \sigma \times \tau$ as in Chapter I. Then, what is $\pi(\sigma,\tau)|_P$? We have seen that $P\backslash G/P$ is finite and only one double coset has positive measure. In fact we can take as a representative of that double coset, the element $m_0' \in M'$ that switches all positive roots to negative roots. Then $P \cap m_0' P m_0' = MAN \cap MAV = MA$, and so

$$\pi(\sigma,\tau)|_P \cong \mathrm{Ind}_{MA}^{P}\, m_0' \cdot (\sigma \times \tau).$$

(2) Again for G semisimple, let P be a cuspidal parabolic, $V = \tilde{\theta}N$. Compute $\pi(\sigma,\tau)|_V$! Aside from a set of measure zero there is only one $P:V$ double coset, and we may take the identity as a representative. Since $P \cap V = \{e\}$ we have

$$\pi(\sigma,\tau)|_V \cong \mathrm{Ind}_{\{e\}}^{V}\, 1_\sigma \cong (\dim \sigma)\lambda_V.$$

(3) Suppose $G = HN$ is a semidirect product, N normal. Let $\gamma \in \hat{N}$, $\pi_\gamma = \mathrm{Ind}_N^G\, \gamma$. What is $\pi_\gamma|_H$? Clearly there is only one $H:N$ double coset, the identity may be taken as a representative element, and

$$\pi_\gamma|_H \cong \mathrm{Ind}_{\{e\}}^{H}\, \gamma|_{\{e\}} \cong (\dim \gamma)\lambda_H.$$

(4) Back to G semisimple. Let P be a cuspidal parabolic and K a maximal compact subgroup. What about $\pi(\sigma,\tau)|_K$? Again since $G = KP$ there is only one double coset, and so

$$\pi(\sigma,\tau)|_K \cong \mathrm{Ind}_{K\cap P}^{K}\, \sigma|_{K\cap P} = \mathrm{Ind}_{K\cap M}^{K}\, \sigma|_{K\cap M}.$$

We can use this equation and the compact reciprocity theorem (see no. 3) to show that if $K \cap M$ is large in M, then K is large in G, at least for the P-series. (Note: $K \cap M$ is the maximal compact of M.) Indeed suppose $\sigma|_{K\cap M} = \sum^{\oplus} n_i \sigma_i$, $\sigma_i \in (K \cap M)^\wedge$, $n_i < \infty$. Then (using the notation $n(\pi,\pi')$ for the number of times a completely reducible representation π contains an irreducible representation π'), we have for $\nu \in \hat{K}$

$$n(\pi(\sigma,\tau)|_K,\nu) = n(\mathrm{Ind}_{K\cap M}^{K}\ \sigma|_{K\cap M},\nu) = \sum n_i n(\mathrm{Ind}_{K\cap M}^{K}\ \sigma_i,\nu)$$

$$= \sum n_i n(\nu|_{K\cap M},\sigma_i) < \infty.$$

This is because the finite-dimensionality of ν guarantees that there are only finitely many σ_i for which $n(\nu|_{K\cap M},\sigma_i) > 0$. That proves the result.

EXERCISES. (1) Suppose G is abelian, and $\pi \in \hat{G}_1$ is a character. Suppose also that G_2 is such that G_1G_2 is a closed subgroup. Show that

$$\mathrm{Ind}_{G_1}^{G}\ \pi|_{G_2} \cong [G:G_1G_2]\mathrm{Ind}_{G_1\cap G_2}^{G_2}\ \pi|_{G_1\cap G_2}.$$

(2) Let Γ be a discrete uniform subgroup of $SL(2,\mathbb{R})$ such that Γ contains no elements of finite order except $\begin{pmatrix} \pm 1 & 0 \\ 0 & \pm 1 \end{pmatrix}$. Show that $\mathrm{Ind}_{\Gamma}^{G}\ 1|_{SO(2)} \cong \infty\lambda$, where $\lambda = \mathrm{Ind}_{Z_G}^{SO(2)}\ 1$.

2. Tensor product theorem. We start with the statement of the theorem.

THEOREM 2. (Tensor Product Theorem - Mackey [3]) *Let* G_1 *and* G_2 *be regularly related subgroups of* G. *Let* π_1 *and* π_2 *be representations of* G_1 *and* G_2 *respectively. For each* $(x,y) \in G \times G$ *consider the subgroup* $G_{x,y} = (x^{-1}G_1x) \cap (y^{-1}G_2y)$ *and the representation of that group* $\pi_{x,y}$ *given by*

$$\pi_{x,y}(g) = \pi_1(xgx^{-1}) \otimes \pi_2(ygy^{-1}).$$

Then the representation

$$\pi^{x,y} = \mathrm{Ind}_{G_{x,y}}^{G}\ \pi_{x,y}$$

is determined to within equivalence by the $G_1:G_2$ *double coset* D *to which* xy^{-1} *belongs, write it* π^D. *Finally, if* ν *is any admissible measure on* $G_1\backslash G/G_2$, *then*

$$\mathrm{Ind}_{G_1}^{G}\ \pi_1 \otimes \mathrm{Ind}_{G_2}^{G}\ \pi_2 \cong \int_{G_1\backslash G/G_2}^{\oplus} \pi^D d\nu(D).$$

Proof. (Sketch) We obtain this result as a consequence of the subgroup theorem in the following way. The representation $\mathrm{Ind}_{G_1}^{G}\ \pi_1 \otimes \mathrm{Ind}_{G_2}^{G}\ \pi_2$ is the representation of G obtained from the representation $\mathrm{Ind}_{G_1}^{G}\ \pi_1 \times \mathrm{Ind}_{G_2}^{G}\ \pi_2$ of $G \times G$ by restricting to the diagonal $\tilde{G} = \{(g,g): g \in G\}$. Noting first that $\mathrm{Ind}_{G_1}^{G}\ \pi_1 \times \mathrm{Ind}_{G_2}^{G}\ \pi_2 \cong \mathrm{Ind}_{G_1\times G_2}^{G\times G}\ \pi_1 \times \pi_2$, and then plugging these ingredients into the subgroup theorem, we obtain the result.

EXAMPLES. (1) For semisimple Lie groups, the various principal series are induced representations -- and so their tensor products should come under the domain of Theorem 2. We will discuss this in detail in section B.

(2) The theorem gives no new information on tensor products of arbitrary representations; that is, if $G_1 = G_2 = G$, the theorem has no content. In general, the problem of decomposing tensor products is extremely difficult.

(3) Take $G_1 = G_2 = \{e\}$, $\pi_1 = \pi_2 =$ the trivial representation. Then $G_1\backslash G/G_2 = G$, $G_{x,y} = \{e\}$, $\pi_{x,y}$ is the trivial representation, and $\pi^{x,y} = \lambda_G$. Therefore

$$\mathrm{Ind}_{\{e\}}^{G}\ 1 \otimes \mathrm{Ind}_{\{e\}}^{G}\ 1 \cong \int_{G}^{\oplus} \lambda_G d\nu(D) = [G:\{e\}]\lambda_G.$$

Hence $\lambda_G \otimes \lambda_G \cong \lambda_G$. The more precise result to the effect that for any $\pi \in \hat{G}$, $\lambda_G \otimes \pi \cong (\dim \pi)\lambda_G$ may be found in Fell [3].

(4) Suppose $G = HN$ and $\nu \in \hat{H}$, $\gamma \in \hat{N}$. Since $H\backslash G/N$ has only one point, we have

$$\mathrm{Ind}_H^G \nu \otimes \mathrm{Ind}_N^G \gamma \cong \mathrm{Ind}_{H\cap N}^G(\nu|_{H\cap N} \otimes \gamma|_{H\cap N}).$$

In particular if $H \cap N = \{e\}$, then

$$\mathrm{Ind}_H^G \nu \otimes \mathrm{Ind}_N^G \gamma \cong (\dim \nu)(\dim \gamma)\lambda_G.$$

(5) Here is an interesting computation which does not seem to be in print anywhere. Let G be a connected semisimple Lie group with finite center, K a maximal compact subgroup, A the split component of a minimal parabolic. For this example we need

LEMMA 3. (Harish-Chandra [3]) $G = KAK$. *Moreover if* $g = k_1 a k_2 = k_3 a' k_4$, *then there is an element* s *of the Weyl group* W *such that* $sa = a'$. *In particular*

$$K\backslash G/K \cong A/W.$$

Now we apply the subgroup and tensor product theorems in this situation. First for $\sigma \in \hat{K}$, we compute $\mathrm{Ind}_K^G \sigma|_K$. Choose a cross-section for the W orbits. In fact if we agree to ignore a set of measure zero, we can take it to be $A^+ = \exp \mathfrak{a}^+$, $\mathfrak{a}^+$ = the positive Weyl chamber = $\{Y \in \mathfrak{a} : \lambda(Y) > 0,\ \lambda \in \Sigma^+\}$. The next problem is to identify the intersection $K \cap a^{-1}Ka$, $a \in A^+$.

EXERCISE. Show that for $a \in A^+$, $k \in K$, then $a^{-1}ka \in K \Longleftrightarrow k \in M$. (Hint: Use the Cartan involution and the fact that $a^2 \in A^+$.) Then we have

$$\mathrm{Ind}_K^G \sigma|_K \cong \int_{A^+}^{\oplus} \mathrm{Ind}_M^K \sigma|_M \, d\nu(a) = \infty\, \mathrm{Ind}_M^K \sigma|_M.$$

We can also apply the tensor product theorem to obtain

$$\mathrm{Ind}_K^G\,\sigma_1 \otimes \mathrm{Ind}_K^G\,\sigma_2 = \int_{A^+}^{\oplus} \mathrm{Ind}_M^G\,\sigma_1|_M \otimes \mathrm{Ind}_M^G\,\sigma_2|_M\,d\nu(a)$$

$$= \infty\cdot\mathrm{Ind}_M^G\,\sigma_1|_M \otimes \mathrm{Ind}_M^G\,\sigma_2|_M.$$

3. Intertwining number theorem - Frobenius reciprocity. In order to determine multiplicities the actual object one usually wants to compute is an intertwining number. It is a difficult matter to compute such numbers generally; e.g., one representation may be weakly contained in another but not actually contained as a subrepresentation, and the intertwining number may be zero. A somewhat more tractable object is the strong intertwining number. An intertwining operator is called a *strong intertwining operator* if it is a Hilbert-Schmidt operator. For representations π_1, π_2 write $\mathcal{J}(\pi_1, \pi_2)$ for the dimension of the space of strong intertwining operators (recall that $\mathcal{I}(\pi_1, \pi_2)$ denotes the dimension of the space of all intertwining operators).

When π is a representation, we let $\mathcal{H}_\pi^f$ be the smallest closed subspace of $\mathcal{H}_\pi$ which contains all finite-dimensional invariant subspaces of π. Set $\pi^f = \pi$ restricted to $\mathcal{H}_\pi^f$.

LEMMA 4. $\mathcal{J}(\pi_1, \pi_2) = \mathcal{I}(\pi_1^f, \pi_2^f)$ = *the number of times that* $\pi_1 \otimes \bar{\pi}_2$ *contains the identity as a discrete direct summand.*

Proof. Let T be any strong intertwining operator for π_1 and π_2. Let $\mathcal{M}_2$ be the orthogonal complement of the null space of T, $\mathcal{M}_1$ the closure of the range of T. Since T intertwines π_1 and π_2, both $\mathcal{M}_1$ and $\mathcal{M}_2$ are invariant. Let $S = T^*T$, a compact operator on $\mathcal{H}_{\pi_1}$. S has pure point spectrum and each eigenvalue occurs with finite multiplicity. It follows that $\mathcal{M}_2$ is a direct sum of finite-dimensional invariant subspaces. Similarly for $\mathcal{M}_1$. Thus $\mathcal{M}_2 = \mathcal{H}_{\pi_1}^f$, $\mathcal{M}_1 = \mathcal{H}_{\pi_2}^f$. Hence every strong intertwining operator carries $\mathcal{H}^f_{\pi_1}$ into $\mathcal{H}^f_{\pi_2}$ and is zero on the orthogonal complement. Therefore $\mathcal{J}(\pi_1,\pi_2) = \mathcal{J}(\pi_1^f, \pi_2^f)$. Finally it is obvious that $\mathcal{J}(\pi_1^f, \pi_2^f) = \mathcal{I}(\pi_1^f, \pi_2^f)$.

EXERCISE. Prove the second equation of Lemma 4.

LEMMA 5. *Let* $\gamma \in \mathrm{Rep}(H)$, *where* $H \subseteq G$ *is a closed subgroup. The number of times that* $\mathrm{Ind}_H^G \gamma$ *contains* 1_G *as a discrete direct summand is equal to the number of times that* γ *contains* 1_H *as a discrete direct summand, provided* G/H *admits a finite invariant measure. If* G/H *does not admit a finite invariant measure, then* $\mathrm{Ind}_H^G \gamma$ *does not contain the identity as a discrete direct summand.*

Proof. See Mackey [3]. It is more or less straightforward by writing down the obvious map.

With these two lemmas and the tensor product theorem, we can prove the strong intertwining number theorem -- at least in the discrete case. For ease of presentation, we confine ourselves to that situation here. For the most general result see Mackey [3].

THEOREM 6. (Strong Intertwining Number Theorem - Mackey [3]) *Let* G_1 *and* G_2 *be discretely related subgroups of* G, $\pi_1 \in \mathrm{Rep}(G_1)$, $\pi_2 \in \mathrm{Rep}(G_2)$. *For* $(x,y) \in G \times G$, *let* $\mathcal{J}(\pi_1,\pi_2;x,y)$ *denote* $\mathcal{J}(s \to \pi_1(xsx^{-1}), s \to \pi_2(ysy^{-1}))$ *considered as representations of* $(x^{-1}G_1x) \cap (y^{-1}G_2y)$. *Then* $\mathcal{J}(\pi_1,\pi_2;x,y)$ *depends only on the* $G_1:G_2$ *double coset to which* xy^{-1} *belongs, write it* $\mathcal{J}(\pi_1,\pi_2,D)$. *Moreover, whether or not* $G/(x^{-1}G_1x \cap y^{-1}G_2y)$ *admits a finite invariant measure depends only on* D. *Let* $\mathcal{D}_f$ *be the set of double cosets for which a finite invariant measure exists and which are not of measure zero. Then*

$$\mathcal{J}(\mathrm{Ind}_{G_1}^G \pi_1, \mathrm{Ind}_{G_2}^G \pi_2) = \sum_{D \in \mathcal{D}_f} \mathcal{J}(\pi_1,\pi_2,D).$$

Proof. By Lemma 4, $\mathcal{J}(\mathrm{Ind}\ \pi_1, \mathrm{Ind}\ \pi_2)$ is equal to the number of times $\mathrm{Ind}\ \pi_1 \otimes \overline{\mathrm{Ind}\ \pi_2} \cong \mathrm{Ind}\ \pi_1 \otimes \mathrm{Ind}\ \bar{\pi}_2$ contains the identity. By Theorem 2, $\mathrm{Ind}\ \pi_1 \otimes \mathrm{Ind}\ \bar{\pi}_2$ is a direct sum over the double cosets of positive measure of certain induced representations $\mathrm{Ind}\ \pi_D$. Hence $\mathcal{J}(\mathrm{Ind}\ \pi_1, \mathrm{Ind}\ \pi_2)$ is the sum over these double cosets of the number of times $\mathrm{Ind}\ \pi_D$ contains the identity. Now by Lemma 5, a given D contributes to this sum only if $D \in \mathcal{D}_f$, and then its contribution is the number of times the Kronecker product π_D contains the identity. But once again by Lemma 4, the number of times π_D contains the identity is exactly $\mathcal{J}(\pi_1, \pi_2, D)$.

COROLLARY. (i) (Finite-Dimensional Reciprocity) *Let* $H \subseteq G$, $\gamma \in \hat{H}$ *and suppose* $\pi \in \hat{G}$ *is finite-dimensional. If* G/H *admits a finite invariant measure, then* $\mathrm{Ind}_H^G\ \gamma$ *contains* π *as a discrete direct summand as many times as* $\pi|_H$ *contains* γ *as a discrete direct summand. If* G/H *has no finite invariant measure,* $\mathrm{Ind}_H^G\ \gamma$ *has no finite-dimensional discrete direct summands.*

(ii) *If* G *is not compact, then its regular representation contains no finite-dimensional discrete direct summands.*

Proof. (i) Take $G_1 = H$, $G_2 = G$ in Theorem 6.

(ii) Take $H = \{e\}$ in (i).

EXAMPLES. (1) If G is a compact group, we get the classical Frobenius Reciprocity Theorem: $H \subseteq G$, $\gamma \in \hat{H}$, $\pi \in \hat{G}$; $\mathrm{Ind}_H^G\ \gamma$ contains π exactly as many times as $\pi|_H$ contains γ.

(2) Let $G = HN$ be a semidirect product with N normal and abelian, and H compact. Let $\chi \in \hat{N}$ be a character, and let $\sigma \in \hat{H}$ be lifted to G. Then $n(\mathrm{Ind}_N^G\ \chi, \sigma) = n(\sigma|_N, \chi) = 0$. So in the decomposition of $\mathrm{Ind}_N^G\ \chi$, none of the finite-dimensional irreducibles of H can occur as direct summands.

(3) Let S be a (solvable) group, N a closed subgroup such

that S/N has finite invariant measure. Let $\chi \in \hat{S}$ be a character, $\gamma \in \hat{N}$ another character. Then

$$n(\mathrm{Ind}_N^S \gamma , \chi) = n(\chi|_N , \gamma) = \begin{cases} 0 & \chi|_N \neq \gamma \\ 1 & \chi|_N = \gamma . \end{cases}$$

It's a sad but true fact that the finiteness conditions of the preceding Corollary are essential to its truth. To see that, consider the following

EXAMPLE. Let G be a connected complex semisimple Lie group, $P = MAN$ a minimal parabolic. Then we know that $\mathrm{Ind}_{MAN}^G 1$ is irreducible. If the Corollary were true in complete generality, then taking $H = P$, $\gamma = 1$, $\pi = \mathrm{Ind}_P^G 1$, it should be true that $\pi|_P$ contains the identity as a discrete direct summand exactly once. But we computed $\pi|_P$ on pp. 47-48. We saw that

$$\pi|_P \cong \mathrm{Ind}_{MA}^{MAN} 1 = \rho .$$

Claim: ρ does not contain the identity. For if so there would exist a non-zero $f \in L_2(N)$ satisfying

$$\delta^{\frac{1}{2}}(a) f(m^{-1}a^{-1}\nu man) = f(\nu), \quad \text{a.a. } \nu \in N$$

for all $m \in M$, $a \in A$, $n \in N$. (Here δ is defined by $\delta(a)\int_N \phi(a^{-1}\nu a)d\nu = \int_N \phi(\nu)d\nu$.) But setting $m = a = e$, we see that f is constant a.e. on N, and hence not in L_2.

4. <u>Mackey and Anh reciprocity</u>. We have already seen that the intertwining number theorem gives a good reciprocity theorem for finite-dimensional representations (in particular for compact groups), but that for infinite-dimensional representations it breaks down. The goal in this no. is to extend the reciprocity as best as possible to

infinite-dimensional representations. We give two results -- Mackey's result from [4], and a recent result due to Anh [1].

THEOREM 7. (Mackey Reciprocity - Mackey [4]) *Suppose* $H \subseteq G$ *and both groups have type* I *regular representations:* $\lambda_G = \int_{\hat{G}}^{\oplus} (\dim \pi)\, \pi \, d\mu_G(\pi)$, $\lambda_H = \int_{\hat{H}}^{\oplus} (\dim \gamma)\, \gamma \, d\mu_H(\gamma)$. *Then there exists a Borel measure* α *on* $\hat{G} \times \hat{H}$ *and an* α*-measurable function* $(\pi,\gamma) \to n(\pi,\gamma)$ *from* $\hat{G} \times \hat{H}$ *to the countable cardinals such that for all Borel sets* S,T *of* $\hat{G},\hat{H}$ *respectively,* $\alpha(S \times \hat{H}) = \mu_G(S)$, $\alpha(\hat{G} \times T) = \mu_H(T)$ *and*

for μ_H*- almost all* $\gamma \in \hat{H}$ $\qquad \operatorname{Ind}_H^G \gamma = \int_{\hat{G}}^{\oplus} n(\pi,\gamma)\, \pi \, d\alpha_\gamma(\pi)$,

for μ_G*- almost all* $\pi \in \hat{G}$ $\qquad \pi|_H = \int_{\hat{H}}^{\oplus} n(\pi,\gamma)\, \gamma \, d\alpha_\pi(\gamma)$,

where $\alpha_\gamma, \alpha_\pi$ *are the quotient measures obtained from* α *by the equivalence relations* $r(\pi,\gamma) = \gamma$, $r(\pi,\gamma) = \pi$.

Proof. (In case G is finite) This proof is not classical -- it is due to Mackey. It is this proof that admits a generalization to the infinite case.

Let J be the identity representation of the diagonal $\tilde{G}$ of $G \times G$. Let $V = \operatorname{Ind}_{\tilde{G}}^{G\times G} J|_{H\times G}$. It is easily shown that $V \cong \operatorname{Ind}_{\tilde{H}}^{H\times G} I$, I being the identity representation of the diagonal $\tilde{H} \subseteq H \times H$. Set $K = \operatorname{Ind}_{\tilde{H}}^{H\times H} I$. Then $V = \operatorname{Ind}_{H\times H}^{H\times G} K$. Next one proves that

$$\operatorname{Ind}_{\tilde{G}}^{G\times G} J = \sum\nolimits^{\oplus} \bar{M} \times M : \ M \in \hat{G}$$

$$K = \sum\nolimits^{\oplus} \bar{L} \times L : \ L \in \hat{H}.$$

Therefore $V = \sum^{\oplus} \bar{L} \otimes \operatorname{Ind}_H^G L = \sum^{\oplus} \bar{M}|_H \otimes M$. Now set

$$\operatorname{Ind} L = \sum n(L,M) M \qquad M|_H = \sum k(M,L) L.$$

Then equating coefficients of $\bar{L} \times M$ in the two decompositions of V, we find $k(M,L) = n(L,M)$ for all L and M. That establishes the reciprocity.

PROBLEMS. (1) Is α equivalent to the product measure $\mu_G \times \mu_H$. The answer is not clear, but it would appear that it doesn't have to be.

(2) The formulas in the theorem are valid only up to non-uniquely specified sets of measure zero. Thus given a specific representation $\pi \in \hat{G}$, it is impossible to say anything definitive about $n(\pi,\gamma)$.

Examples are given in Mackey [4]. They are cumbersome and can be illustrated better with the aid of the following improved version due to Anh.

THEOREM 8. (Anh Reciprocity - Anh [1]) *Let* $H \subsetneqq G$, *both type* I *groups. Let* μ_G, μ_H *be finite measures in the class determined by the regular representations of* G,H *respectively. Let* $\omega(\pi,\gamma)$ *and* $n(\pi,\gamma)$ *be* $\mu_G \times \mu_H$*-measurable functions, where* $n(\pi,\gamma)$ *is a countable cardinal for every* $\pi \in \hat{G}$, $\gamma \in \hat{H}$. *Then the following are equivalent*

(i) *for* μ_H*-almost all* $\gamma \in \hat{H}$

$$\mathrm{Ind}_H^G \gamma \cong \int_{\hat{G}}^{\oplus} n(\pi,\gamma)\, \pi \, d\alpha_\gamma'(\pi)$$

where $d\alpha_\gamma'(\pi) = \omega(\pi,\gamma) d\mu_G(\pi)$;

(ii) *for* μ_G*-almost all* $\pi \in \hat{G}$

$$\pi|_H = \int_{\hat{H}}^{\oplus} n(\pi,\gamma)\, \gamma \, d\alpha_\pi'(\gamma)$$

where $d\alpha_\pi'(\gamma) = \omega(\pi,\gamma) d\mu_H(\gamma)$.

The proof proceeds by showing that either (i) or (ii) is

equivalent to the condition that α is absolutely continuous with respect to $\mu_G \times \mu_H$. The function ω is then the Radon-Nikodym derivative.

EXAMPLES. (1) Let G be type I and $H \subseteq G$, a compact group. Let $\pi \in \hat{G}$ be arbitrary. Then $\pi|_H$ is a direct sum of irreducibles

$$\pi|_H = \sum_{\gamma \in \hat{H}}^{\oplus} n(\pi,\gamma)\gamma.$$

It follows by Theorem 8, that for μ_H - almost all $\gamma \in \hat{H}$

$$\mathrm{Ind}_H^G\ \gamma = \int_{\hat{G}}^{\oplus} n(\pi,\gamma)\, \pi\, d\mu_G(\pi).$$

But since H is compact, μ_H is atomic. Therefore for *all* $\gamma \in \hat{H}$, we have

$$\mathrm{Ind}_H^G\ \gamma = \int_{\hat{G}}^{\oplus} n(\pi,\gamma)\, \pi\, d\mu_G(\pi),$$

where the multiplicities $n(\pi,\gamma)$ are uniquely specified by the direct sum decomposition for $\pi|_H$. In particular if H is large in G, then all of the numbers $n(\pi,\gamma)$ are finite. In case G is a connected semisimple Lie group with finite center, $H = K$ a maximal compact subgroup, it is known that there is a constant D such that

$$n(\pi,\gamma) \leqq D \dim \gamma, \qquad \pi \in \hat{G}.$$

Therefore the irreducibles π that occur in $\mathrm{Ind}_H^G\ \gamma$ all occur with multiplicity less than or equal to the fixed number $D \dim \gamma$. Even more specifically, let $\pi \in \hat{G}_d$. Then in $\pi|_K = \sum^{\oplus} n(\pi,\gamma)\gamma$, choose $\gamma \in \hat{K}$ so that $n(\pi,\gamma) > 0$. Then π must occur $n(\pi,\gamma)$ times as a discrete direct summand of $\mathrm{Ind}_K^G\ \gamma$. Of course it is a very difficult problem (as we have seen) to pick out a single irreducible subspace.

(2) $H = \{e\}$. For any $\pi \in \hat{G}$, clearly $\pi|_H = \dim \pi \cdot 1$. Therefore

$$\lambda_G = \text{Ind}_H^G 1 = \int_{\hat{G}}^{\oplus} (\dim \pi)\, \pi\, d\mu_G(\pi),$$

thereby recovering the canonical decomposition of the regular representation into irreducibles, each occuring with multiplicity equal to its dimension.

(3) Let G be a connected semisimple Lie group with finite center. Assume G has only one conjugacy class of Cartan subgroups. Then the Plancherel measure is concentrated in the set of principal series $\pi(\sigma,\tau) = \text{Ind}_P^G\, \sigma \times \tau$ arising from a minimal parabolic P.

(a) We have computed previously, using the subgroup theorem, that

$$\pi(\sigma,\tau)|_K \cong \text{Ind}_M^K\, \sigma.$$

But the representations $\pi(\sigma,\tau)$ constitute a.a. the members of $\hat{G}$. Therefore by Anh reciprocity, for all $\gamma \in \hat{K}$ we have

$$\text{Ind}_K^G\, \gamma \cong \int_{\hat{G}}^{\oplus} n(\sigma,\tau,\gamma)\pi(\sigma,\tau)d\mu_G(\sigma,\tau)$$

where $n(\sigma,\tau,\gamma) = n(\sigma,\gamma)$ = the number of times that γ is contained in $\text{Ind}_M^K\, \sigma$ = the number of times that $\gamma|_M$ contains σ. This is of course a sharper estimate than obtained in (1).

(b) We also computed $\pi(\sigma,\tau)|_V$ previously via the subgroup theorem. In fact

$$\pi(\sigma,\tau)|_V \cong (\dim \sigma)\lambda_V.$$

Once again, the representations $\pi(\sigma,\tau)$ constitute a.a. of $\hat{G}$; thus by Anh reciprocity, for μ_V - almost all $\kappa \in \hat{V}$ we have

$$\text{Ind}_V^G\, \kappa \cong \int_{\hat{G}}^{\oplus} (\dim \sigma)(\dim \kappa)\pi(\sigma,\tau)d\mu_G(\sigma,\tau).$$

Lately (mostly because of Jacquet and Langlands [1]) people have become interested in realizations of representations of G as subrepresentations of induced representations from V (so-called *Whittaker models*). One is especially interested in the above when the multiplicity is one. For that, clearly one needs $(\dim \sigma)(\dim \kappa) = 1$; in which case M and V must both be abelian. That essentially forces $G = SL(2,\mathbb{R})$ or $SL(2,\mathbb{C})$. Therefore in more general cases (V is nilpotent), it has been necessary to consider characters of V -- representations which are of measure zero in $\hat{V}$ (see Chapter IV) -- and so for which the reciprocity theorem is not helpful.

(4) Finally we give examples to show the limitations of Anh reciprocity. Let $H \subsetneqq G$ be co-compact, i.e., G/H compact. Assume also that H is CCR. Now suppose we are interested in computing the multiplicities in the representations $\pi|_H$, $\pi \in \hat{G}$. Then to use Anh reciprocity we look at $\mathrm{Ind}_H^G \gamma$. But the hypotheses guarantee that $\mathrm{Ind}_H^G \gamma$ is a discrete direct sum of irreducibles (with finite multiplicity). This is useless, since if G is not compact the counting measure on $\hat{G}$ is almost never absolutely continuous with respect to Plancherel measure.

Here is another case. Let G be a connected semisimple Lie group, $H = P$ a cuspidal parabolic. One is interested in computing $\pi(\sigma,\tau)|_P$. In principal this is possible by the subgroup theorem (see e.g. Example (1), p. 47). In reality, an explicit description of $\pi(\sigma,\tau)|_P$ decomposed into irreducibles seems to be a difficult problem. Another method would be to compute $\mathrm{Ind}_P^G \gamma$ for a.a. representations $\gamma \in \hat{P}$ and then use Anh reciprocity. But that requires knowledge of the Plancherel theory of P, another difficult and unsolved problem. This appears to be a fertile area for further research.

Finally, Anh reciprocity is of no use in the situation of discrete subgroups $\Gamma \subsetneqq G$ since these groups are rarely type I.

B. APPLICATIONS TO SEMISIMPLE GROUPS

The main question that we consider here is the following: G is a connected semisimple Lie group with finite center, P is a cuspidal parabolic; decompose into irreducibles the tensor product $\pi(\sigma_1,\tau_1) \otimes \pi(\sigma_2,\tau_2)$ of two representations in the P-series. Williams [1] first solved this problem when G is a complex group. Martin [1] obtained significant results in case G has $\mathbb{R}$-rank one. We give an outline of a general procedure for attacking the problem, and of the specific results in the two mentioned cases.

First we wish to apply the tensor product theorem. Let $\bar{P}$ be the parabolic opposed to P (i.e., if $P = MAN$, then $\bar{P} = MA\bar{N}$, $\bar{N} = \tilde{\theta}N$). Now $P\backslash G/\bar{P}$ breaks up into a point and a set of measure zero. We take the identity as the representative of the coset which has positive measure. Then by Chapter I, Lemma C 2 (iii), and the tensor product theorem, we have

$$\begin{aligned}\pi(\sigma_1,\tau_2) \otimes \pi(\sigma_2,\tau_2) &\cong \mathrm{Ind}_P^G\ \sigma_1 \times \tau_1 \otimes \mathrm{Ind}_{\bar{P}}^G\ \sigma_2 \times \tau_2 \\ &\cong \mathrm{Ind}_{P\cap\bar{P}}^G\ (\sigma_1 \times \tau_1) \otimes (\sigma_2 \times \tau_2) \\ &= \mathrm{Ind}_{MA}^G\ (\sigma_1 \otimes \sigma_2) \times \tau_1\tau_2 .\end{aligned}$$

Hence the problem reduces to knowing how to decompose tensor products of discrete series of M, and representations of the form $\mathrm{Ind}_{MA}^G\ \sigma \times \tau$, $\sigma \in \hat{M}_d$, $\tau \in \hat{A}$. Very little is known about the former when M is not compact; so we assume that P is minimal. Then M is compact and we can write

$$\sigma_1 \times \sigma_2 = \sum_{\hat{M}}^{\oplus} n_i \sigma_i$$

the n_i being the so-called Clebsch-Gordan coefficients. In terms of these the problem is then reduced to decomposing

$$\mathrm{Ind}_{MA}^{G}\ \sigma \times \tau, \quad \sigma \in \hat{M}, \quad \tau \in \hat{A}.$$

We can effect a further reduction of the problem with the following result.

THEOREM 1. (Martin [1]) *Let* $\sigma \in \hat{M}$, $\tau, \tau' \in \hat{A}$. *Then*

$$\mathrm{Ind}_{MA}^{G}\ \sigma \times \tau \cong \mathrm{Ind}_{MA}^{G}\ \sigma \times \tau',$$

that is, $\mathrm{Ind}_{MA}^{G}\ \sigma \times \tau$ *is (up to unitary equivalence) independent of* $\tau \in \hat{A}$.

Proof. (Sketch) Let α be a *simple* restricted root, i.e., a positive root that is not a sum of other positive roots. Let $\alpha_1, \ldots, \alpha_\ell$ be a set of simple roots. Any positive root α may be written $\alpha = \sum m_i \alpha_i$, $m_i \in \mathbb{Z}_+$. Note $\ell = \dim A = \mathbb{R}$-rank of G. Let $\alpha = \alpha_i$, $\mathfrak{n}_\alpha = \sum_{k \geq 1} \mathfrak{g}_{k\alpha}$, $\mathfrak{v}_\alpha = \theta\, \mathfrak{n}_\alpha$, $\tilde{\mathfrak{g}}^\alpha$ = the subalgebra of $\mathfrak{g}_0$ generated by $[\mathfrak{n}_\alpha, \mathfrak{v}_\alpha]$. Then $\tilde{\mathfrak{g}}^\alpha = \mathbb{R}H_\alpha + \mathfrak{m}_\alpha$ where $\mathfrak{m}_\alpha = H_\alpha^\perp$ in $\tilde{\mathfrak{g}}^\alpha$, and $\mathfrak{g}^\alpha = \tilde{\mathfrak{g}}^\alpha + \mathfrak{n}_\alpha + \mathfrak{v}_\alpha$ is an $\mathbb{R}$-rank one semisimple subalgebra of $\mathfrak{g}$. Using the fact that $\tau(\exp Y) = \exp\left(\sum_i \tau_i \alpha_i(Y)\right)$, $Y \in \mathfrak{a}$, $\tau_i \in \mathbb{R}$ and the above constructions, it is possible to reduce to the case of $\mathbb{R}$-rank one groups, i.e., $\mathfrak{g} = \mathfrak{g}^\alpha$.

Next using the theorem on induction in stages, we see that it is enough to prove $\mathrm{Ind}_{MA}^{P}\ \sigma \times \tau \cong \mathrm{Ind}_{MA}^{P}\ \sigma \times \tau'$. We then use the exponential map and the definition of induced representations to realize these representations in $L_2(\mathfrak{n}\,; \mathcal{H}_\sigma)$. The proof progresses by taking the abelian Fourier transform on $\mathfrak{g}_{2\alpha}$ (if $\mathfrak{g}_{2\alpha} \neq \{0\}$) or $\mathfrak{g}_\alpha$ (if $\mathfrak{g}_{2\alpha} = \{0\}$). The resulting representations are then seen to be equivalent via a suitable multiplication operator (see Martin [1] for the details.

It follows from Theorem 1 that once we know $\mathrm{Ind}_{MA}^{G}\ \sigma \times \tau$ for μ_{MA}-a.a. $\sigma \times \tau$, we actually know $\mathrm{Ind}_{MA}^{G}\ \sigma \times \tau$ for all $\sigma \in \hat{M}$, $\tau \in \hat{A}$.

In order to learn what $\mathrm{Ind}_{MA}^{G}\ \sigma\times\tau$ is a.e. we apply Anh reciprocity. So we have to consider $\pi|_{MA}$ for irreducible representations π of G. First consider the principal series corresponding to a minimal parabolic P, $\pi(\sigma,\tau) = \mathrm{Ind}_{P}^{G}\ \sigma\times\tau$. Since we want to decompose $\pi(\sigma,\tau)|_{MA}$, it is natural to try the subgroup theorem. Thus we need to discuss MAN: MA double cosets.

LEMMA 2. *Let* MA *act on* $V = \tilde{\theta}N$ *by inner automorphism of* G. *Suppose that* S *is a cross-section for* V/MA. *Then* S *will also serve as a cross-section for the double coset space* MAN\G/MA *with the exception of a set of measure zero.*

Proof. Consider the map

$$\phi: V/MA \to MAN\backslash G/MA, \qquad \phi\cdot\mathcal{O}_v \to MANvMA,$$

$\mathcal{O}_v$ = an orbit containing $v \in V$. It is easy to check that ϕ is a 1 - 1 Borel isomorphism of V/MA onto MAN\MANV/MA. But MANV is a co-null, dense open submanifold of G, and the lemma follows.

REMARK. If S is a Borel set, this implies that MA and MAN are regularly related.

Now in case G is complex (resp. G is $\mathbb{R}$-rank one), then Williams [1] (resp. Martin [1]) has constructed a cross-section S as specified in Lemma 2. It is essentially given as follows. Let Σ' denote the set of positive roots which are not simple roots. In the complex case there are dim V - dim A elements in Σ'; in the $\mathbb{R}$-rank one case there is 0 or 1 according as $\mathfrak{n}_{2\alpha}$ is trivial or not. Choose $X_\alpha \in \mathfrak{n}_\alpha$, $X_\alpha \neq 0$, $\alpha \in \Sigma'$. Then it is only a minor innacuracy to state that

$$S = \{\exp(\sum_{\alpha\in\Sigma'} y_\alpha X_\alpha): y_\alpha \in \mathbb{R}^* \}$$

is a cross-section for V/MA.

Now for $s \in S$, let $(MA)_s$ denote the stability group for s, $(MA)_s = \{ma \in MA: masm^{-1}a^{-1} = s\}$.

LEMMA 3. *The cross-section may be chosen so that* $(MA)_s = M_s$ *is independent of* $s \in S$; *denote it* M_0.

The proof is more or less by computation. When G is complex $M_0 = Z_G$; when G has $\mathbb{R}$-rank one, M_0 is a "computable" compact Lie subgroup of M.

Next note that $sMANs^{-1} \cap MA = M_0$. This follows from the fact that $V \to MAN\backslash G$, $v \to MANv$, is an MA-equivalent map. The general principal is that if H_2 acts on $H_1\backslash G$ on the right, then the stability group of $H_1 g \in H_1\backslash G$ is $H_2 \cap g^{-1}H_1 g$.

We can now apply the subgroup theorem to arrive at the following result

$$\begin{aligned}\pi(\sigma,\tau)|_{MA} &= (\text{card } S)\ \text{Ind}_{M_0}^{MA}\ \sigma|_{M_0} \\ &= (\text{card } S)(\text{Ind}_{M_0}^{M}\ \sigma|_{M_0} \times \lambda_A) \\ &= (\text{card } S)(\sum_{\nu\in\hat{M}} n(\sigma,\nu)\nu \times \lambda_A)\end{aligned}$$

where $n(\sigma,\nu) = \mathcal{I}(\sigma|_{M_0}, \nu|_{M_0})$. If G is complex, then $\text{card } S = \infty$ unless $G = SL(2,\mathbb{C})$ in which case S consists of a single point. If G has $\mathbb{R}$-rank one, then $\text{card } S = \infty$ unless $G = SO_e(n,1)$. In that case $\text{card } S = 1$, $n \geqq 3$ (actually it's 2 if $n = 2$ -- that is, related to the minor discrepancy).

Now suppose we are in the case of a complex group. Then the representations $\pi(\sigma,\tau)$ constitute a.a. $\pi \in \hat{G}$. The conclusion by Anh reciprocity is that for a.a. $\sigma \times \tau \in (MA)^\wedge$ -- and so all $\sigma \times \tau \in (MA)^\wedge$ by Theorem 1 -- we have

$$\text{Ind}_{MA}^{G}\ \sigma \times \tau \cong \int_{\hat{G}}^{\oplus} (\text{card } S)n(\sigma,\sigma_1)\pi(\sigma_1,\tau_1)d\mu_G(\sigma_1,\tau_1)$$

where $n(\sigma,\sigma_1) = \mathscr{I}(\sigma|_{M_0}, \sigma_1|_{M_0})$. That is Williams' result [1].

Next suppose we are in the $\mathbb{R}$-rank one case. Then the representations $\pi(\sigma,\tau)$ constitute a.a. of $\hat{G}$ only if the non-compact part of G is locally isomorphic to $SO_e(2n+1,1)$, $n \geqq 1$. Otherwise there will be a discrete series as well. Thus we have to say something about restrictions of discrete series representations to MA. Here Martin is only able to compute an estimate rather than a precise multiplicity. By an observation in section A4, if $\pi \in \hat{G}_d$ then there is $\nu \in \hat{K}$ (K = a maximal compact) such that π is a direct summand of $\mathrm{Ind}_K^G \nu$. Let us then look at $\mathrm{Ind}_K^G \nu|_{MA}$. The technique is again via the subgroup theorem. This time we need the $K: MA$ double cosets. Using an Iwasawa decomposition $G = KAV$, we see in a similar fashion to previous computations that it suffices to consider V/M. Roughly, this equals $A\cdot V/MA = A\cdot S$. The result is

$$\mathrm{Ind}_K^G \nu|_{MA} = \infty\ \mathrm{Ind}_{M_0}^{MA} \nu|_{M_0} = \infty\Big(\sum_{\sigma\in\hat{M}} n(\nu,\sigma)\sigma \times \lambda_A\Big)$$

where $n(\nu,\sigma) = \mathscr{I}(\nu|_{M_0}, \sigma|_{M_0})$. Thus we obtain that for $\pi \in \hat{G}_d$, $n(\pi,\nu) > 0$,

$$\pi|_{MA} = \int_{(MA)^\wedge}^{\oplus} n(\pi,\sigma)(\sigma \times \tau)d\mu_{MA}(\sigma,\tau)$$

where $n(\pi,\sigma) \leqq \infty n(\nu,\sigma)$.

In any event we have accounted for a.a. the irreducibles of G. Thus we are justified in using Anh reciprocity. The final result is

$$\mathrm{Ind}_{MA}^G \sigma \times \tau = \int^{\oplus} (\mathrm{card}\ S)n(\sigma,\sigma_1)\pi(\sigma_1,\tau_1)d\mu_G(\sigma_1,\tau_1)$$

$$\oplus \sum_{\pi\in\hat{G}_d} n(\pi,\sigma)\pi.$$

The multiplicities in the continuous spectra are explicitly determined;

in the discrete spectra, we can only say which representations cannot occur. For more information the reader is referred to Martin [1].

PROBLEMS. (1) Compute $n(\pi,\sigma)$ explicitly for $\pi \in \hat{G}_d$.

(2) Drop the assumption $\mathbb{R}$-rank one. That involves considering other cuspidal parabolics and in general looking at

$$\mathrm{Ind}_P^G \sigma \times \tau|_{M_1 A_1}$$

where $P = MAN$, $P_1 = M_1A_1N_1$ are two cuspidal parabolics. This seems to be a difficult problem and progress has been slow.

(3) Compute the decomposition of tensor products of discrete series representations. This also seems to be a very difficult problem. Apparently the only substantial work on it is found in Pukanszky [1].

CHAPTER III. REPRESENTATIONS OF GROUP EXTENSIONS

The general idea that we pursue in this chapter is the following: G is a locally compact group, $N \subseteq G$ is a closed normal subgroup; we try to describe the representation theory of G in terms of the representation theories of N and G/N. In this way one hopes to reduce matters to questions of "lower-dimensional" groups, and eventually to groups without normal subgroups, e.g., the semisimple Lie groups which we have discussed in Chapter I. Well, the idea has had some measure of success, although perhaps not as much as one might have hoped for.

We give two cases, a preliminary case involving abelian groups and then the most general case. For both, we shall have need of Mackey's Imprimitivity Theorem which is discussed briefly in the second appendix.

A. SEMIDIRECT PRODUCTS WITH ABELIAN GROUPS

The discussion that follows is based on Mackey's treatment in [3]. Let H and N be locally compact groups with N abelian. Suppose there is a homomorphism $H \to \mathrm{Aut}(N)$ such that the map $H \times N \to N$, $(h,n) \to h\cdot n$, is continuous. This enables us to form the semidirect product $G = H\cdot N$. Clearly any unitary representation $(x,y) \to U_{(x,y)}$ of G is determined by its restrictions to H, N. If $x \to V_x$, $y \to W_y$ denote those restrictions, then $U_{(x,y)} = V_x W_y$, $x \in H$, $y \in N$. Conversely let $x \to V_x$, $y \to W_y$ be representations of H,N acting in the same space. Then $(x,y) \to V_x W_y$ is a representation of G if and only if $V_x W_y V_{x^{-1}} = W_{x\cdot y}$. Now by Stone's Theorem, the representation $y \to W_y$ of the locally compact abelian group N is determined by a projection-valued measure $E \to P_E$ on the Borel sets $\mathcal{B}(\hat{N})$

$$W_y = \int_{\hat{N}} \gamma(y) dP(\gamma), \quad y \in N.$$

One checks that V and W satisfy the above identity $\Longleftrightarrow$ P and V satisfy

$$V_x P_E V_{x^{-1}} = P_{x\cdot E}, \quad x \in H, \quad E \in \mathcal{B}(\hat{N}).$$

But the latter is precisely the condition for P to be a system of imprimitivity for V.

Now consider the action of H on $\hat{N}$, $(x\cdot\gamma)(n) = \gamma(x^{-1}\cdot n)$. Suppose the measure P is concentrated in one of the orbits under the action. Then let γ be any member of the orbit and set H_γ = the subgroup of H that stabilizes γ. Then $x \to x\cdot\gamma$, $H/H_\gamma \to H\cdot\gamma$, is a Borel isomorphism. In this way P becomes a system of imprimitivity for V based on H/H_γ. By the imprimitivity theorem, we may conclude that

$$V = \mathrm{Ind}_{H_\gamma}^{H} \nu$$

for some $\nu \in \mathrm{Rep}(H_\gamma)$. Then one shows that under reasonable conditions (specified below), the representation U cannot be irreducible unless P is in fact concentrated in an orbit.

DEFINITION. G is called a *regular semidirect product* of H and N if $\hat{N}$ contains a countable family $E_1, E_2, \ldots$ of Borel sets, each a union of orbits such that every orbit is the intersection of the E_j which contain it.

THEOREM 1. *Let* G *be a regular semidirect product of* H *and* N, *with* N *abelian and normal. For each orbit* $H\cdot\gamma$, $\gamma \in \hat{N}$, *set* $H_\gamma = \{h \in H: h\cdot\gamma = \gamma\}$. *Let* $U_{(x,y)} = V_x W_y \in \hat{G}$. *Then the projection-valued measure defined by* W *in* $\hat{N}$ *is concentrated in a single orbit* $H\cdot\gamma$

and $V = \mathrm{Ind}_{H}^{H} \nu$ *for some* $\nu \in \hat{H}_\gamma$. *Every pair consisting of an orbit* $H\cdot\gamma$ *and an irreducible* ν *arise in this way, and two of these are equivalent* $\Longleftrightarrow$ *the orbits are identical and the representations* ν *are equivalent.*

Here is an alternate (and better for the purposes of generalization) form of the above representations. Let $H\cdot\gamma$ be an orbit, H_γ = the stability group, and take $\nu \in \hat{H}_\gamma$. Then the corresponding representation U may be written $U = \mathrm{Ind}_{H_\gamma N}^{G} \nu\otimes\gamma$, $(\nu\otimes\gamma)(hn) = \nu(h)\gamma(n)$. That this is so can be verified by the subgroup theorem. Indeed, one needs to compute $\mathrm{Ind}_{H_\gamma N}^{G} \nu\otimes\gamma|_H$ and $\mathrm{Ind}_{H_\gamma N}^{G} \nu\otimes\gamma|_N$. But $H_\gamma N\backslash G/H$ has one point and so

$$\mathrm{Ind}_{H_\gamma N}^{G} \nu\otimes\gamma|_H = \mathrm{Ind}_{H_\gamma}^{H} \nu.$$

On the other hand $H_\gamma N\backslash G/N \cong H_\gamma\backslash H$, and so $H_\gamma N$ and N are regularly related. Moreover one checks easily (using the subgroup theorem) that $\mathrm{Ind}_{H_\gamma N}^{G} \nu\otimes\gamma|_N$ is a multiple of the direct integral

$$\int_{H/H_\gamma}^{\oplus} h\cdot\gamma \, d\bar{h} \,;$$

and so the resulting projection-valued measure is concentrated in that orbit.

COROLLARY. *If* H *is also abelian, every irreducible representation of* $G = H\cdot N$ *is monomial, that is, induced from a character.*

EXAMPLES. The first four of the following are taken directly from Mackey [3]. The remaining are examples that have been considered by the author and others on various occasions.

(1) In $G = H\cdot N$, let $N = \mathbb{R}$, $H = \mathbb{R}_+^*$ where H acts on N by $a\cdot n = an$, $n \in \mathbb{R}$, $A \in \mathbb{R}_+^*$. Then $\hat{N} \cong \mathbb{R}$ in the usual way and the action of H on $\hat{N}$ is the same as on N. Hence there are three

orbits: $\mathcal{O}_1 = (-\infty,0)$, $\mathcal{O}_2 = \{0\}$, $\mathcal{O}_3 = (0,\infty)$. The stability groups are $H_{-1} = \{e\}$, $H_0 = H$, $H_1 = \{e\}$. Thus the irreducible representations of G are: those trivial on N, i.e., the characters of $\mathbb{R}_+^*$, $a \to a^{it}$, $t \in \mathbb{R}$, lifted to G; and two infinite-dimensional representations

$$\mathrm{Ind}_N^G \gamma_\varepsilon, \quad \gamma_\varepsilon(n) = e^{i\varepsilon n}, \quad n \in \mathbb{R}, \quad \varepsilon = \pm 1.$$

The reader may find it instructive to see what happens if $\mathbb{R}_+^*$ is replaced by $\mathbb{R}^*$.

(2) Let $G = H\cdot N$, $N = \mathbb{C}$, $H = \mathbb{T}$, and let H act on N by $z\cdot x = zx$, $z \in \mathbb{T}$, $x \in N$. Of course $\hat{N} \cong \mathbb{C}$, and the action of $\mathbb{T}$ is unchanged. The orbits are circles centered at the origin

$$\mathcal{O}_\rho = \{\gamma \in \hat{N}: |\gamma| = \rho\}, \quad \rho \geqq 0.$$

Also $H_\rho = \begin{cases} \mathbb{T} & \rho = 0 \\ \{e\} & \rho > 0 \end{cases}$. Thus the irreducible representations of G fall into two classes: representations trivial on N, i.e., characters of $\mathbb{T}$, $z \to z^m$, $m \in \mathbb{Z}$, lifted to G; and the family

$$\pi_\rho = \mathrm{Ind}_N^G \rho, \quad \rho > 0.$$

(3) $G = H\cdot N$, $N = \mathbb{R}^2$, $H = \left\{\begin{pmatrix} a & b \\ 0 & a^{-1}\end{pmatrix} : a > 0, \ b \in \mathbb{R}\right\}$, and H acts on N by matrix multiplication $(x,y)\cdot\begin{pmatrix} a & b \\ 0 & a^{-1}\end{pmatrix} = (ax, bx + a^{-1}y)$. Then $\hat{N} \cong \mathbb{R}^2$ via $\gamma(x,y) = e^{i(\gamma_1 x + \gamma_2 y)}$ say, and $h^{-1}\cdot\gamma = (a\gamma_1 + b\gamma_2, a^{-1}\gamma)$ if $h = \begin{pmatrix} a & b \\ 0 & a^{-1}\end{pmatrix}$. Thus there are five orbits: $\mathcal{O}_1 = (0,\infty)$, $\mathcal{O}_2 = (-\infty,0)$, $\mathcal{O}_3 = \{0\}$, $\mathcal{O}_4$ = the upper half-plane, $\mathcal{O}_5$ = the lower half-plane. The corresponding stability groups are: $H_{(1,0)} = H_{(-1,0)} = U = \left\{\begin{pmatrix} 1 & b \\ 0 & 1\end{pmatrix} : b \in \mathbb{R}\right\}$, $H_{(0,0)} = H$, $H_{(0,1)} = H_{(0,-1)} = \{e\}$. Corresponding to the orbits $\mathcal{O}_1$, $\mathcal{O}_2$ we get two

one-parameter family of representations induced from $U \cdot N$; to the orbit $\mathcal{O}_3$, we get the representations of example 1; and to the orbits $\mathcal{O}_4$, $\mathcal{O}_5$ we get two representations induced from N itself.

(4) $G = H \cdot N$, $N = \mathbb{C}$, $H = \mathbb{Z}$, where for a fixed irrational number λ we let H act on N by $m \cdot z = e^{2\pi i \lambda m} z$, $m \in \mathbb{Z}$, $z \in \mathbb{C}$. The action on $\hat{N}$ is the same. Each orbit is countable and dense in a circle and of course there are uncountably many such on each circle. It follows easily that $G = H \cdot N$ is not a regular semidirect product.

(5) $G = H \cdot N$, $N = F^n$, $H = SL(n,F)$, F a locally compact field of characteristic zero and H acts on N by matrix multiplication. We shall discuss local fields in more detail in Chapter V; for now, the uninitiated reader may assume $F = \mathbb{R}$. The dual $\hat{N} \cong F^n$ and the group action is essentially the same. Clearly $\hat{N}/G$ consists of two points $\{0\}$ and $F^n - \{0\}$. The corresponding stabilizers are $H = SL(n,F)$ and $H_1 = \{h \in SL(n,F)\colon h \cdot (1,0,\ldots,0) = (1,0,\ldots,0)\}$. It's straightforward to compute that

$$H_1 = \begin{pmatrix} 1 & * \cdots * \\ 0 & \\ \vdots & * \\ 0 & \end{pmatrix} \cong SL(n-1,F) \cdot F^{n-1}.$$

If $n = 2$, the latter is just $\left\{ \begin{pmatrix} 1 & b \\ 0 & 1 \end{pmatrix} : b \in F \right\} \cong F$. By a simple induction argument it is easy to see that $\hat{G} = \mathcal{S}_1 \cup \cdots \cup \mathcal{S}_n$ where $\mathcal{S}_j \approx SL(j,F)^\wedge$, $2 \leqq j \leqq n$, and $\mathcal{S}_1 \approx \hat{F}$.

(6) **EXERCISE.** Take $G = H \cdot N$, $N = \mathbb{Q}_p$ = the field of p-adic numbers, H = the group of units in $\mathbb{Q}_p$ and let H act on N by multiplication. Compute $\hat{G}$. Show that it is countable.

(7) G is called a *motion group* if $G = H \cdot N$, where N is normal and abelian and H is compact. Such groups must be regular (see

later comments in section B), and so their representation theory is given by Theorem 1. Hence, roughly speaking, there are two parameters -- one for the orbits $H\cdot\gamma$, $\gamma \in \hat{N}$, and the other for the duals $\hat{H}_\gamma$ of the stability groups. We specialize to two important subcases.

(a) *Euclidean motion groups.* Here take $N = \mathbb{R}^n$, $H = SO(n)$ and H is acting on N or $\hat{N}$ by rotations. The orbits are then spheres centered at the origin, say

$$\mathcal{O}_\rho = \{\gamma \in \hat{N}: \|\gamma\| = \rho\}, \quad \rho \geqq 0.$$

Then $H_\rho = \begin{cases} H & \rho = 0 \\ SO(n-1) & \rho > 0 \end{cases}$. Hence the representations fall into two classes: those trivial on N, i.e., finite-dimensional representations of H lifted to G; and the two parameter family

$$\pi_{\sigma,\rho} = \mathrm{Ind}_{SO(n-1)\cdot N}^{G}\ \sigma\rho, \quad \sigma \in SO(n-1)^\wedge, \quad \rho \in \hat{N}, \quad \rho = (\rho,0,\ldots,0), \quad \rho > 0.$$

(b) *Cartan motion groups.* Take G_1 to be a connected semisimple Lie group with finite center, $\mathfrak{g}_1 = \mathfrak{k} + \mathfrak{p}$ a Cartan decomposition, K a corresponding maximal compact subgroup. Then restricting the adjoint representation of G_1 to the subgroup K and the subspace $\mathfrak{p}$, we see that $\mathfrak{p}$ is a K-module. Let $G = H\cdot N$ where N is the vector group corresponding to the $\mathbb{R}$-linear space structure of $\mathfrak{p}$ and $H = K$. H acts on N by the module action. It is known that a fundamental domain for this action is the closure $\overline{\mathfrak{a}^+}$ of a positive Weyl chamber $\mathfrak{a}^+$. For $Y \in \mathfrak{a}^+$, $K_Y = M$ $(= Z(\mathfrak{a}) \cap K)$. For other $Y \in \overline{\mathfrak{a}^+} - \mathfrak{a}^+$, the stability groups K_Y are larger than M in general. They can be classified according to the number of walls Y is on. Next identify N to $\hat{N}$ via the Killing form, which is positive definite and K-invariant on $\mathfrak{p}$. More precisely, $Y \to \gamma_Y$ is chosen so that $\gamma_Y(Y_1) = e^{2\pi i B(Y,Y_1)}$,

$Y_1 \in \mathcal{P}$. Then $K_{\gamma_Y} = K_Y$, $Y \in \mathcal{P}$. Thus aside from a "small set" the representations of G form a two-parameter family

$$\pi_{\sigma,Y} = \mathrm{Ind}_{MN}^{G}\ \sigma\gamma_Y, \quad \sigma \in \hat{M}, \quad Y \in \mathfrak{a}^+.$$

(8) *Inhomogeneous Lorentz groups.* Let $H = SO_e(n,1)$ as defined previously. Since H is a subgroup of $GL(n+1,\mathbb{R})$ it acts on $\mathbb{R}^{n+1}$ by matrix multiplication and we may form $G = H\cdot N$, $N = \mathbb{R}^{n+1}$. We can identify N with $\hat{N}$ via the non-degenerate form $x_1y_1 + \ldots + x_ny_n - x_{n+1}y_{n+1}$. The action of H on $\hat{N}$ is then the same as on N, and the orbit picture on $\hat{N}$ is a familiar one:

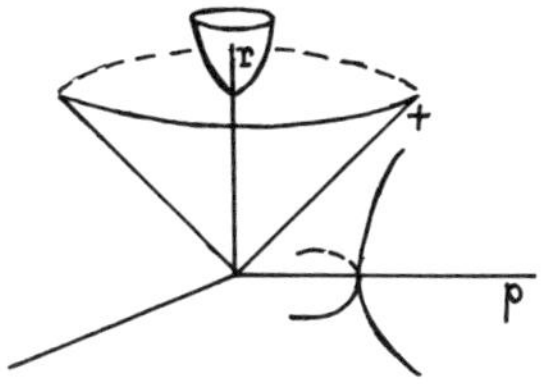

There are four kinds of orbits: $\mathcal{O}_0 = \{0\}$, $\mathcal{O}_\pm$ = the (light) cones, $\mathcal{O}_{\pm r}$ = the hyperboloids of rotation, $\mathcal{O}_\rho$ = the hyperboloids of revolution. The corresponding stability groups are: $H_0 = H$, $H_+ = MU$, $H_- = M\tilde{\theta}U$ (where MAU is a minimal parabolic), $H_{\pm r} = K$ (maximal compact subgroup), $H_\rho = SO_e(n-1,1)$. The reader is referred to Kleppner and Lipsman [1,2] for more details on this example. We also leave it as an exercise to write down the various parameters that describe the elements of $\hat{G}$.

(9) As an abstraction of many of these examples, we can consider the following. Let H be a real or complex Lie group, V a vector space over the reals or complexes accordingly. Suppose $\sigma: H \to GL(V)$ is a finite-dimensional representation. Form the semidirect product $G = H\cdot\hat{V}$ and inquire: (a) when is G a regular semidirect product; and (b) what are the ingredients of the Mackey theory? This is a highly non-trivial problem which is more properly studied from the

point of view of algebraic groups. We return to it later in Chapter V.

(10) The following is due to Mautner and Mackey (see Mackey [7]). Consider $N = \mathbb{C}^2$, λ a fixed irrational number, and $H = \mathbb{R}$. Form $G = H \cdot N$ where H acts on N by $x \cdot (z,w) = (e^{2\pi ix}z, e^{2\pi i\lambda x}w)$. This product is not regular and G is not type I. Next let $\psi: G \to H$ denote the canonical projection. Then we define an action of G on the reals by $g \cdot u = e^{\psi(g)}u$, and consider the semidirect product $G \cdot \mathbb{R}$. There are three orbits in $\mathbb{R}$: $(-\infty,0)$, $\{0\}$, $(0,\infty)$. The stability group corresponding to the first or third is $\{g \in G: \psi(g) = 0\} = \mathbb{C}^2$. The stability group corresponding to the second orbit is G itself. Thus the fiber over the "small orbit" $\{0\}$ is $\hat{G}$, the dual of a non-type I group; while the fiber over the others is the smooth space $\hat{\mathbb{C}}^2$. So we can parameterize "most" of the representations by two copies of $\hat{\mathbb{C}}^2$. One conclusion that can be drawn (see Mackey [7] and Kleppner and Lipsman [2]) is that although $G \cdot \mathbb{R}$ is not type I, its regular representation is type I.

B. THE GENERAL CASE

The situation we wish to consider now is $N \subsetneqq G$ closed and normal, but we don't assume that N is abelian or that it splits in G. How can we carry over the Mackey method developed in section A? Well, we try as far as possible to proceed as in the abelian case. Let π be an irreducible representation of G. We wish to associate with $\pi|_N$ a projection-valued measure on $\hat{N}$. For this we need our first assumption, namely that N is type I. In particular then $\hat{N}$ is a standard Borel space. Now $\pi|_N$ is a type I representation and hence is quasi-equivalent to a well-determined multiplicity-free representation π^N of N. In fact if $\pi|_N = \int_{\hat{N}}^{\oplus} m_\pi(\gamma)\, \gamma \, d\mu_\pi(\gamma)$, then $\pi^N = \int_{\hat{N}}^{\oplus} \gamma \, d\mu_\pi(\gamma)$ where the class of μ_π is uniquely determined. For each

Borel set $E \subseteq \hat{N}$, let $\{\mu_\pi^E\}$ denote the measure class given by $\mu_\pi^E(F) = \mu_\pi(F \cap E)$, $F \in \mathcal{B}(\hat{N})$. Then $\pi^E = \int_{\hat{N}}^{\oplus} \gamma \, d\mu_\pi^E(\gamma)$ is a subrepresentation of π^N. Let P_E be the unique projection in the center of $\mathcal{R}(\pi^N, \pi^N)$ determined by this subrepresentation. Roughly speaking, P_E is multiplication by the characteristic function of E. In any event, it is the projection-valued measure we are after.

Next G acts on $\hat{N}$ in a canonical way: $(g\cdot\gamma)(n) = \gamma(g^{-1}ng)$, $n \in N$, $g \in G$, $\gamma \in \hat{N}$. One then verifies that

$$\pi_y P_E \pi_y^{-1} = P_{y.E}, \quad y \in G, \quad E \in \mathcal{B}(\hat{N}).$$

In short, P is a system of imprimitivity for π.

Now suppose that P is concentrated in an orbit of G in $\hat{N}$. Let γ be in the orbit and let G_γ be the stability group. The map $g \to g\cdot\gamma$, $G/G_\gamma \to G\cdot\gamma$, is then a Borel isomorphism. Thus P becomes a system of imprimitivity for π based on G/G_γ. By the imprimitivity theorem, we can conclude that

$$\pi = \operatorname{Ind}_{G_\gamma}^{G} \nu$$

for some $\nu \in \operatorname{Rep}(G_\gamma)$ (necessarily irreducible if π is irreducible).

Next we have to give a sufficient condition for all the projection-valued measures P to be concentrated in an orbit.

DEFINITION. We say that N is *regularly embedded* in G if the space $\hat{N}/G$ with the quotient Borel structure is countably separated, i.e., if there is a countable family $E_1, E_2, \ldots$ of Borel sets, each a union of orbits such that every orbit is the intersection of the E_j which contain it.

LEMMA 1. (Mackey [6]) *If* N *is regularly embedded in* G, *then every* P *arising from* $\pi \in \hat{G}$ *as above is concentrated in an orbit.*

Next we observe that the representations ν which arose (in the realization $\pi = \mathrm{Ind}_{G_\gamma}^{G} \nu$) are not arbitrary irreducibles. In fact an application of the Subgroup Theorem reveals the following: since N is normal $N\backslash G/G_\gamma = G/G_\gamma$ and

$$\mathrm{Ind}_{G_\gamma}^{G} \nu|_N = \int_{G/G_\gamma}^{\oplus} x\cdot\nu|_N \, d\bar{g}.$$

But we also have, setting $\pi = \mathrm{Ind}_{G_\gamma}^{G} \nu$

$$\pi|_N = \int_{\hat{N}}^{\oplus} m_\pi(\gamma_1)\gamma_1 d\mu_\pi(\gamma_1).$$

It follows by a closer examination of the projection-valued measures implicit in these two decompositions (the latter of which is P of course) that for almost all $x \in G$, $x\cdot\nu|_N = m(\gamma)(x\cdot\gamma)$ (see Mackey [6]). In particular $\nu|_N$ is a multiple of γ.

In light of these facts we can now state a theorem.

THEOREM 2. (Mackey [6]) *Let* $N \subseteq G$ *be a* type I, *regularly embedded subgroup. Choose* $\gamma \in \hat{N}$, *one from each* G-*orbit, and let* G_γ *be the stability group. Let* $\check{G}_\gamma = \{\nu \in \hat{G}_\gamma : \nu|_N$ is a multiple of $\gamma\}$. *Then for* $\nu \in \check{G}_\gamma$, $\pi_\nu = \mathrm{Ind}_{G_\gamma}^{G} \nu$ *is irreducible and* $\hat{G}$ *is a disjoint union*

$$\hat{G} = \bigcup_{\hat{N}/G} \{\pi_\nu : \nu \in \check{G}_\gamma\}.$$

The next step is to give a finer description of the elements of $\check{G}_\gamma$. Recall that in the semidirect product $G = H\cdot N$ (with N abelian), we had $G_\gamma = H_\gamma N$ and any $\nu \in \check{G}_\gamma$ was of the form $\nu = \sigma \otimes \gamma$, $\nu(hn) = \sigma(h)\gamma(n)$, $h \in H_\gamma$, $n \in N$, $\sigma \in \hat{H}_\gamma$. We are not so fortunate in the general situation. But suppose that γ extends to a representation γ' of G_γ. Clearly if $\sigma \in (G_\gamma/N)\hat{}$, $\sigma'' \in \hat{G}_\gamma$ its lift, then $\nu = \sigma'' \otimes \gamma' \in \check{G}_\gamma$. Conversely, suppose $\nu \in \check{G}_\gamma$. Then $\nu|_N$ is a multiple of γ. Hence there exists a Hilbert space $\mathcal{H}_0$ such that $\mathcal{H}_\nu = \mathcal{H}_0 \otimes \mathcal{H}_\gamma$ and $\nu(n) = I \otimes \gamma(n)$. Then for $n \in N$, $x \in G$,

$$\nu(xnx^{-1}) = I \otimes \gamma(xnx^{-1}) = (I \otimes \gamma'(x))(I \otimes \gamma(n))(I \otimes \gamma'(x)^{-1})$$

and

$$\nu(xnx^{-1}) = \nu(x)\nu(n)\nu(x)^{-1} = \nu(x)(I \otimes \gamma(n))\nu(x)^{-1}.$$

It follows that $\nu(x)^{-1}(I \otimes \gamma'(x))$ commutes with $I \otimes \gamma(n)$. Since γ is irreducible, we have

$$\nu(x)^{-1}(I \otimes \gamma'(x)) = T(x) \otimes I$$

where $T(x)$ is a unitary operator on $\mathcal{H}_0$. Note that $T(nx) \otimes I = \nu(nx)^{-1}(I \otimes \gamma'(nx)) = \nu(x)^{-1}(I \otimes \gamma'(x)) = T(x) \otimes I$. Setting $\sigma(\bar{x}) = T(x)^{-1}$, $\bar{x} \in G_\gamma/N$, we see that $\nu = \sigma'' \otimes \gamma'$. It follows by the irreducibility of ν that σ must be irreducible.

Unfortunately in general γ cannot be extended to a representation of G_γ -- there is an obstruction. Indeed, the extension if it exists should be obtained as follows: $x \in G_\gamma \Longrightarrow x\cdot\gamma$ and γ are equivalent; hence there is a unitary T such that $T\gamma(x^{-1}nx) = \gamma(n)T$. By the irreducibility of γ, T is uniquely determined up to scalars. Question: can we choose $T = \gamma'(x)$ so that $x \to \gamma'(x)$ will be a representation? Answer: the best you can do in general is a projective or multiplier representation.

For the reader's benefit we include here a brief discussion of multiplier representations. Indeed, by a *multiplier representation* π of G, we mean a map $\pi\colon G \to \mathcal{U}(\mathcal{H})$ such that $\pi(e) = I$, $\pi(x)\pi(y) = \omega(x,y)\pi(xy)$, and $x \to (\pi(x)\xi,\eta)$ is a Borel function for all $\xi,\eta \in \mathcal{H}$. The function ω is called a *multiplier* and π is an ω-representation. ω is a Borel function on $G \times G$ satisfying $|\omega| = 1$, $\omega(e,x) = \omega(x,e) = 1$ and $\omega(xy,z)\omega(x,y) = \omega(x,yz)\omega(y,z)$. One can define the usual notions for ω-representations: irreducibility, equivalence, factor representation, intertwining operator, etc. We say (G,ω) is type I if all the ω-representations generate type I von-Neumann algebras. Denote by $\hat{G}^{\omega}$ the equivalence classes of irreducible ω-representations of G.

If ρ is a Borel function on G, then $\omega_{\rho}(x,y) = \rho(x)\rho(y)/\rho(xy)$ is a *trivial* multiplier. ω and ω_1 are called *similar* if $\omega = \omega_1\omega_{\rho}$ for some ρ. It's easy to see that ω and ω_1 have isomorphic representation theories. Clearly the product of two multipliers is a multiplier; in fact the set $\mathcal{M}(G)$ of multipliers forms an abelian group with $\bar{\omega} = \omega^{-1}$. The collection $\mathcal{T}(G)$ of trivial multipliers is a subgroup and the quotient $\mathcal{M}(G)/\mathcal{T}(G)$ is usually called the *multiplier group* of G. Note of course that if π_1 and π_2 are ω_1- and ω_2-representations respectively, then $\pi_1 \otimes \pi_2$ is an $\omega_1\omega_2$-representation.

Finally we indicate how the ω-representation theory of G can be subsumed under a portion of the ordinary representation theory of a group extension of G. Set $G(\omega) = \mathbb{T} \times G$ with multiplication

$$(t,x)(s,y) = (ts\omega(x,y),xy), \quad t,s \in \mathbb{T}, \quad x,y \in G.$$

If dt,dx denote right Haar measures on $\mathbb{T},G$, then it is easy to check that $dt\,dx$ is a right invariant measure on $G(\omega)$. In fact $G(\omega)$ has a unique locally compact topology (not the product topology in general) such that: $dt\,dx$ is a right Haar measure, $\mathbb{T}$ is a closed central subgroup, and $G(\omega)/\mathbb{T} \cong G$. If π is an ω-representation of G, then $\pi^0(t,x) = t\pi(x)$, $t \in \mathbb{T}$, $x \in G$, is an ordinary representation of $G(\omega)$. In fact the map $\pi \to \pi^0$ is an injection of $\hat{G}^{\omega}$ into $\{\pi^0 \in \hat{G}(\omega): \pi^0(t,x) = t\pi^0(1,x)\}$. For more information on these matters, see Auslander and Moore [1] or Kleppner and Lipsman [1,2].

EXAMPLE. Take $G = \mathbb{R}^2$, $\omega_{\gamma}((x,y),(\xi,\eta)) = e^{i\gamma(\xi y - x\eta)/2}$, $\gamma \in \mathbb{R}^*$. ω_{γ} is a multiplier on G. If we take $\mathcal{H} = L_2(G)$, then we can define a representation

$$\sigma_\gamma(x,y)f(u) = e^{-i\gamma y(u+\frac{1}{2}x)}f(u+x), \quad f \in \mathcal{H}.$$

This is an irreducible ω_γ-representation of G, and in fact it is the only one up to equivalence (Mackey [6]).

Now we continue with Mackey's description of $\check{G}_\gamma$.

THEOREM 3. (Mackey [6]) *Let* $N \subseteq H$ *be closed and normal,* $\gamma \in \hat{N}$, $H_\gamma = H$. *Then there exists a multiplier* ω *(uniquely determined up to similarity) on* H/N *such that if* ω *also denotes the lift to* H, *then* γ *extends to an* ω*-representation* γ' *of* H *and any* $\nu \in \check{H}$ *is of the form* $\nu = \sigma'' \otimes \gamma'$, *where* $\sigma \in (\widehat{H/N})^{\overline{\omega}}$, σ'' *is the lift to* H. *The class of* ν *is given uniquely by the class of* σ.

Combining the previous results, we obtain the following fundamental result about representations of group extensions.

THEOREM 4 (Mackey [6]) *Let* $N \subseteq G$ *be a* type I *regularly embedded subgroup. Then* $\hat{G}$ *is a fiber set with* $\hat{N}/G$ *as base and fibers* $(\widehat{G_\gamma/N})^{\overline{\omega}_\gamma}$, $\pi = \pi_{\gamma,\sigma} = \mathrm{Ind}_{G_\gamma}^{G}\sigma'' \otimes \gamma'$. *Moreover* G *is* type I *if and only if all the pairs* $(G_\gamma/N, \overline{\omega}_\gamma)$ *are* type I.

EXAMPLES. (1) Of course all the previous examples in section A illustrate Theorem 4 as well as Theorem A1.

Let us now see several examples in which N is not abelian or not split.

(2) Let $N \subseteq G$ be type I and suppose G/N is compact. It's automatic then that N is regularly embedded in G. One way of seeing that is the following. For any $\gamma \in \hat{N}$, the map $G/G_\gamma \to G\cdot\gamma$ is continuous. By the compactness of G/G_γ, it must be a homeomorphism. It then follows from one of Glimm's equivalent conditions (Glimm [1])

that $\hat{N}/G$ is countably separated. Also the fibers in this case are actually discrete -- since $(G_\gamma/N)(\bar{\omega}_\gamma)$ is a compact group. In particular G must be type I. By being more specific in several different ways, we can obtain some special cases:

(a) G central, that is G/Z_G compact. These groups have been studied in considerable detail by Grosser and Moskowitz [1].

(b) Groups all of whose irreducible representations are finite-dimensional. Moore [3] derived a structure theorem for such groups -- essentially they must be a projective limit of finite extensions of central groups. One can then in principle write down the representation theory for such groups (see also Lipsman [4]).

(c) $G = GL(n,F)$, F a locally compact field of characteristic zero. Then $[F^*: F^{*^n}] < \infty$ (see the remark on p.108). Set $G_n = \{g \in GL(n,F): \det g \in F^{*^n}\}$, a closed normal subgroup of finite index in G. But the map $(g_{ij}) \times a \to (ag_{ij})$, $SL(n,F) \times F^* \to G_n$ is a continuous and open homomorphism onto G_n with finite kernel. With this observation, Theorem 4 provides in principle the means of computing $\hat{G}$ if one knows $SL(n,F)^\wedge$.

(3) *The Heisenberg group.* Let $G = \left\{\begin{pmatrix} 1 & x & z \\ 0 & 1 & y \\ 0 & 0 & 1 \end{pmatrix} : x,y,z \in \mathbb{R}\right\}$, $N = \left\{\begin{pmatrix} 1 & 0 & z \\ 0 & 1 & 0 \\ 0 & 0 & 1 \end{pmatrix} : z \in \mathbb{R}\right\} = Z_G$. We take $\hat{N} = \{\gamma: z \to e^{i\gamma z},\ \gamma \in \mathbb{R}\}$. Then $G_\gamma = G$, $\gamma \in \hat{N}$, since N is central. For the extension data, we get

$$\gamma'(x,y,z) = e^{i\gamma(z-\frac{1}{2}xy)}$$

$$\omega_\gamma[(x,y),(\xi,\eta)] = e^{i\gamma(\xi y-x\eta)/2}.$$

Taking $\bar{\sigma}_\gamma = \sigma_{-\gamma}$ as on pp. 78-79, we obtain a one-parameter family of representations

$$\pi_\gamma = \pi_{\gamma,\sigma_{-\gamma}} = \bar{\sigma}_\gamma \otimes \gamma'$$

$$\pi_\gamma(x,y,z)f(u) = e^{i\gamma(z+yu)}f(u+x), \quad f \in L_2(\mathbb{R}).$$

We leave it to the reader to check that π_γ is the unique irreducible representation π of N such that $\pi(z) = e^{i\gamma z}$, $z \in N$.

(4) Let $G = H\cdot N$ where N is the Heisenberg group of example (3), $H = SL(2,\mathbb{R})$ and H acts on N as follows. It fixes the center Z of N and it moves the other two variables according to the matrix action of $SL(2,\mathbb{R})\cdot\mathbb{R}^2$ (see Kleppner and Lipsman [1]). Pictorially, $\hat{N}$ looks like this 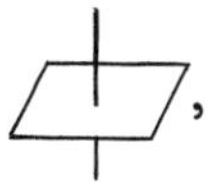, and $\hat{N}/H$ looks like $\rho_1 \,|\, \rho_2$; that is, H (since it leaves Z alone) cannot change the class of the irreducible $\pi_\gamma \in \hat{N}$, $\gamma \in \mathbb{R}^*$, and it shrinks the plane of characters to two points ρ_1,ρ_2. Over ρ_1, the stability group is H itself and we get the representations of $SL(2,\mathbb{R})$ lifted to G. Over ρ_2, the stability group is $\left\{\begin{pmatrix}1 & z\\ 0 & 1\end{pmatrix} : z \in \mathbb{R}\right\}$ and we get a one-parameter family of representations. Actually the latter two families form precisely the representations of G trivial on the normal subgroup Z, i.e., the representations of $G/Z \cong SL(2,\mathbb{R})\cdot\mathbb{R}^2$. The stability group corresponding to π_γ, $\gamma \neq 0$, is $H = SL(2,\mathbb{R})$. But there is an obstruction to extending π_γ to H. This obstruction was first computed by Shale [1] (and later in analogous but more general situations by Duflo [1]). It goes as follows. Take $\tilde{H}$ to be a two-fold covering of H, $p\colon \tilde{H} \to H$ the canonical projection, and $\psi\colon H \to \tilde{H}$ a Borel cross-section. Then there is a unitary representation Π_γ of $\tilde{H}$ such that $\pi'_\gamma = \Pi_\gamma \circ \psi$ is the extension of π_γ. Π_γ is called the *Weil representation*. The fiber over $\{\pi_\gamma\}$ is $\hat{\tilde{H}}' =$
$\{\pi \in (\tilde{H})^\wedge : \pi|_{p^{-1}(\{e\})} \neq 1\}$.

(5) **EXERCISE.** Take the group $G = H\cdot N$ of example (4) and let $H_1 = SO(2) \subseteq H$. Consider the group $G_1 = H_1\cdot N$ (whose simply connected covering is usually called the *oscillator group*). Compute the representation theory of G.

(6) We give one final example (due to Mackey [6]) just to indicate how the ingredients of the extension procedure may become quite complicated, even in relatively low dimension. Let G be the group of 3×3 real lower triangular matrices of determinant 1,

$$G = \left\{ \begin{pmatrix} \lambda & 0 & 0 \\ b & \mu & 0 \\ a & c & \nu \end{pmatrix} : \text{real entries}, \quad \lambda\mu\nu = 1 \right\}.$$

Set

$$N = \left\{ \begin{pmatrix} 1 & 0 & 0 \\ b & 1 & 0 \\ a & c & 1 \end{pmatrix} : a,b,c \in \mathbb{R} \right\} \quad \text{and} \quad D = \left\{ \begin{pmatrix} \lambda & 0 & 0 \\ 0 & \mu & 0 \\ 0 & 0 & \nu \end{pmatrix} : \lambda\mu\nu = 1 \right\}.$$

Clearly G is a semidirect product, $G = D\cdot N$, N normal. We already know the irreducible representations of N: the infinite-dimensional representations π_γ, $\gamma \in \mathbb{R}^*$, and the two-parameter family $\chi_{s,t}$ of characters. Now the action of D on N is given by the equations

$$(\lambda,\mu,\nu)\cdot(a,b,c) = \left(\frac{\nu}{\lambda} a \, , \, \frac{\mu}{\lambda} b \, , \, \frac{\nu}{\mu} c \right).$$

As for the orbits in $\hat{N}$, those are as follows. For the characters $\chi_{s,t}$, we have $\mathcal{O}_1 = \{\chi_{s,t}: s \neq 0,\ t \neq 0\}$, $\mathcal{O}_2 = \{\chi_{s,0}: s \neq 0\}$, $\mathcal{O}_3 = \{\chi_{0,t}: t \neq 0\}$, $\mathcal{O}_4 = \{\chi_{0,0}\}$. The corresponding stability groups are $D_{(1,1)} = \{e\}$, $D_{(1,0)} = (\lambda,\lambda,\lambda^{-2})$, $D_{(0,1)} = (\lambda^{-2},\lambda,\lambda)$, $D_{(0,0)} = D$. So we get several families of representations: one irreducible representation induced from N to G by $\chi_{1,1}$, two "one-parameter" families over the orbits $\mathcal{O}_2$, $\mathcal{O}_3$, and a two-parameter family $\approx \hat{D}$.

Regarding the representations π_γ of N, we see the following. Since π_γ is uniquely determined by its restriction to $Z = \left\{\begin{pmatrix}1 & 0 & 0\\ 0 & 1 & 0\\ a & 0 & 1\end{pmatrix} : a \in \mathbb{R}\right\}$ and $(\lambda,\mu,\nu)\cdot(a,0,0) = (\frac{\nu}{\lambda}a\,,0,0)$, it is obvious that all the representations $\{\pi_\gamma : \gamma \neq 0\}$ constitute a single orbit $\mathcal{O}_5$. The stability group is clearly $D' = (\lambda,\lambda^{-2},\lambda)$. The group D' carries no multipliers, and so at last we get a one-parameter family of representations induced from $D'\cdot N$.

The last topic we discuss in this chapter is the Plancherel formula for group extensions. The reader is referred to section C of Chapter I for the relevant preliminaries on the Plancherel measure for unimodular type I groups. We also remark that if G is not unimodular, it is possible to be more precise about the Plancherel measure μ_G (other than it is one of a class of measures that admits the decomposition $\lambda_G = \int_{\hat{G}}^{\oplus} (\dim \pi)\, \pi \, d\mu_G(\pi)$). We do not discuss that here -- the interested reader is referred to Kleppner and Lipsman [1,2].

Put succinctly, our goal now is to do (for group extensions) to μ_G what we did to $\hat{G}$, namely, try to describe it in terms of corresponding objects for N and G_γ/N. First of all, since multiplier representations of G_γ/N enter into the picture, one needs a "projective Plancherel theorem". In fact, that is not hard to obtain (see Kleppner and Lipsman [1,2]): if (G,ω) is type I, then there is a measure (unique if G is unimodular) $\mu_{G,\omega}$ such that

$$\lambda_{G,\omega} = \int_{\hat{G}^{\omega}}^{\oplus} (\dim \pi)\, \pi \, d\mu_{G,\omega}(\pi)$$

where $\lambda_{G,\omega}$ is the left regular ω-representation defined by

$$\lambda_{G,\omega}(x)f(y) = \Delta_G(x)^{-\frac{1}{2}}\omega(y^{-1},x)f(x^{-1}y), \quad f \in L_2(G).$$

For more information on $\lambda_{G,\omega}$ and related matters see Kleppner and

Lipsman [1,2]. The following result can also be found there (in [1]).

THEOREM 5. *Let* G *be a locally compact group,* X *a standard Borel space,* μ *a quasi-invariant* σ *finite measure on* X. *Let* $\bar{\mu}$ *be a pseudo-image of* μ *on* X/G *and suppose* $\bar{\mu}$ *is countably separated. Then for* μ*-almost all* $x \in X$, *there exists a quasi-invariant measure* μ_x, *concentrated in* $G\cdot x$ *such that*

$$\int_X f(x)d\mu(x) = \int_{X/G}\int_{G/G_x} f(g\cdot x)d\mu_x(\bar{g})d\bar{\mu}(\bar{x}), \quad f \in L_1(X,\mu).$$

By a pseudo-image $\bar{\mu}$ of μ we mean the following: Let ν be a finite measure equivalent to μ and set $\bar{\mu} = p\nu$, $p\colon X \to X/G$, i.e., $\bar{\mu}(E) = \nu(p^{-1}(E))$, $E \in \mathcal{B}(X/G)$. $\bar{\mu}$ is uniquely determined up to equivalence. Finally, to say that $\bar{\mu}$ is countably separated means that after removing a $\bar{\mu}$-null set, X/G is countably separated.

Now using the fact that $\hat{N}$ satisfies the assumptions of Theorem 5, we make a formal computation. First

$$\lambda_N = \int_{\hat{N}}^{\oplus} (\dim\gamma)\,\gamma\, d\mu_N(\gamma) = \int_{\hat{N}/G}^{\oplus}\int_{G/G_\gamma}^{\oplus} \dim\gamma\,(g\cdot\gamma)d\mu_\gamma(\bar{g})d\bar{\mu}_N(\bar{\gamma}).$$

Next induce both sides to G

$$\lambda_G = \mathrm{Ind}_N^G\,\lambda_N = \int_{\hat{N}/G}^{\oplus}\int_{G/G_\gamma}^{\oplus} \mathrm{Ind}_N^G(\dim\gamma\,(g\cdot\gamma))d\mu_\gamma(\bar{g})d\bar{\mu}_N(\bar{\gamma}).$$

Then continue

$$\lambda_G = \int_{\hat{N}/G}^{\oplus} \dim\gamma \int_{G/G_\gamma}^{\oplus} g\cdot\mathrm{Ind}_N^G\,\gamma\; d\mu_\gamma(\bar{g})d\bar{\mu}_N(\bar{\gamma})$$

$$= \int_{\hat{N}/G}^{\oplus} \dim\gamma\,[G:G_\gamma]\mathrm{Ind}_N^G\,\gamma\, d\bar{\mu}_N(\bar{\gamma})$$

$$= \int_{\hat{N}/G}^{\oplus} \dim\gamma\,[G:G_\gamma]\mathrm{Ind}_{G_\gamma}^G\,\mathrm{Ind}_N^{G_\gamma}\,\gamma\, d\bar{\mu}_N(\bar{\gamma})$$

$$= \int_{\hat{N}/G}^{\oplus} \dim\gamma\,[G:G_\gamma]\mathrm{Ind}_{G_\gamma}^{G}(\lambda''_{G_\gamma/N,\bar{\omega}_\gamma} \otimes \gamma')d\bar{\mu}_N(\bar{\gamma})$$

$$= \int_{\hat{N}/G}^{\oplus} \dim\gamma\,[G:G_\gamma]\mathrm{Ind}_{G_\gamma}^{G}\left(\int_{(G_\gamma\hat{/}N)^{\bar{\omega}_\gamma}}^{\oplus} (\dim\sigma)\sigma'' d\mu_{G_\gamma/N,\bar{\omega}_\gamma}(\sigma) \otimes \gamma'\right) d\bar{\mu}_N(\bar{\gamma})$$

$$= \int_{\hat{N}/G}^{\oplus} \int_{(G_\gamma\hat{/}N)^{\bar{\omega}_\gamma}}^{\oplus} \dim\gamma\,[G:G_\gamma]\dim\sigma\ \mathrm{Ind}_{G_\gamma}^{G}(\sigma''\otimes\gamma')d\mu_{G_\gamma/N,\bar{\omega}_\gamma}(\sigma)d\bar{\mu}_N(\bar{\gamma}).$$

Finally observing that $\dim\pi_{\gamma,\sigma} = \dim\gamma\,[G:G_\gamma]\dim\sigma$, $\pi_{\gamma,\sigma} = \mathrm{Ind}_{G_\gamma}^{G}\ \sigma''\otimes\gamma'$, and writing $\mu_\gamma = \mu_{G_\gamma/N,\bar{\omega}_\gamma}$, we find that

$$\lambda_G = \int_{\hat{N}/G}^{\oplus} \int_{(G_\gamma\hat{/}N)^{\bar{\omega}_\gamma}}^{\oplus} (\dim\pi_{\gamma,\sigma})\pi_{\gamma,\sigma} d\mu_\gamma(\sigma)d\bar{\mu}_N(\bar{\gamma}).$$

The above calculation can be made rigorous (see Kleppner and Lipsman [1]), and that enables us to state the following theorem.

THEOREM 6. (Kleppner and Lipsman [1,2]) *Let* $N \subseteq G$ *be* type I *and regularly embedded. Let the pairs* $(G_\gamma/N,\bar{\omega}_\gamma)$ *be* type I (*so that* G *is also* type I). *Then the Plancherel measure class of* G *is a fiber measure in this set up. On the base it is a pseudo-image of Plancherel measure on* $\hat{N}$, *on the fibers it is the projective Plancherel measure of* $(G_\gamma/N,\bar{\omega}_\gamma)$. *If in addition* G *is unimodular, the precise measure* μ_G *is determined as follows: let* $\bar{\mu}_N$ *be any pseudo-image of* μ_N *on* $\hat{N}/G$; *then for* μ_N-a.a. γ, *the projective Plancherel measure* $\mu_\gamma = \mu_{G_\gamma/N,\bar{\omega}_\gamma}$ *is uniquely specified so that*

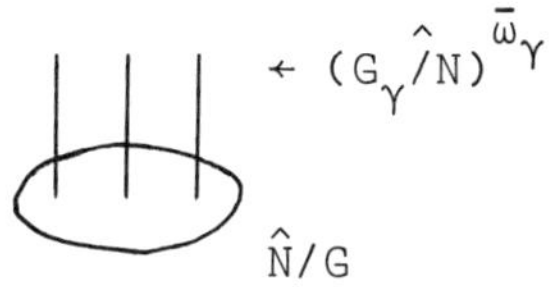

$$\mu_G = \int_{\hat{N}/G} \mu_\gamma\, d\bar{\mu}_N(\gamma),$$

that is, if $\bar{\mu}_N$ *is changed into* $c(\bar{\gamma})d\bar{\mu}_N(\bar{\gamma})$, *then a.a. the fiber measures must be altered by* $c(\bar{\gamma})^{-1}$. *Finally if* G/N *is compact, one may take the image of* μ_N *rather than a pseudo-image.*

EXAMPLE. $G = H\cdot N$, $H = SO_e(n,1)$, $N = \mathbb{R}^{n+1}$. Referring to Example (8) of section A, we see that there are four classes of representations corresponding to the four different types of orbits in $\hat{N}$: $\mathcal{O}_0$, $\mathcal{O}_\pm$, $\mathcal{O}_{\pm r}$, $\mathcal{O}_\rho$. The first two are of measure zero in $\hat{N}/H$. The latter two give representations induced by the parameters $(\pm r,\sigma)$, $r > 0$, $\sigma \in SO(n)^\wedge$ and (ρ,τ), $\rho > 0$, $\tau \in SO_e(n-1,1)^\wedge$ respectively. The Plancherel formula is:

$$\int_G |f(g)|^2 dg = c_1 \int_0^\infty \sum_{\sigma\in SO(n)^\wedge} (\|\pi_{+r,\sigma}(f)\|_2^2 + \|\pi_{-r,\sigma}(f)\|_2^2)\dim\sigma\, r^n dr$$

$$+ c_2 \int_0^\infty \int_{SO_e(n-1,1)^\wedge} \|\pi_{\tau,\rho}(f)\|_2^2 \, d\mu_{SO_e(n-1,1)}(\tau)\rho^n d\rho,$$

(see Kleppner and Lipsman [2]).

EXERCISE. Compute μ_G for all the remaining examples enumerated in sections A and B. A great many of them can be found in Kleppner and Lipsman [1,2].

CHAPTER IV. NILPOTENT GROUPS

In this chapter we give an introduction to the representation theory of simply connected nilpotent Lie groups. The funadmental ideas are due to Kirillov [1]. Our treatment through Theorem 5 follows Quint [1]. Most of the remaining material is taken directly from Kirillov [1]. Because of that we always write our induced actions on the left in this chapter.

A. THE ORBIT THEORY OF KIRILLOV

Let $\mathfrak{g}$ be a finite-dimensional Lie algebra over $\mathbb{R}$, $\mathfrak{g}^* = \mathrm{Hom}_{\mathbb{R}}(\mathfrak{g},\mathbb{R})$, $f \in \mathfrak{g}^*$. Denote by B_f the alternating bilinear form on $\mathfrak{g}$ given by $B_f(x,y) = f([x,y])$, $x,y \in \mathfrak{g}$.

DEFINITION. A subalgebra $\mathfrak{h} \subseteq \mathfrak{g}$ is said to be a *real polarization* at f if $\mathfrak{h}$ is a maximal totally isotropic subspace for B_f (i.e. $f[x,\mathfrak{h}] = 0 \iff x \in \mathfrak{h}$).

We shall see later that $\dim \mathfrak{h} = \frac{1}{2}(\dim \mathfrak{g} + \dim \mathfrak{g}(f))$, $\mathfrak{g}(f) = \{x \in \mathfrak{g} : f([x,\mathfrak{g}]) = 0\}$. Kirillov [1] calls a sublagebra $\mathfrak{h}$ such that $f([\mathfrak{h},\mathfrak{h}]) = 0$, *subordinate* to f. For the nilpotent case, maximal subordinate subalgebras are the real polarizations, but in general that need not be so.

EXAMPLE. Let $\mathfrak{g} = \mathfrak{h}_1 = \mathbb{R}P+\mathbb{R}Q+\mathbb{R}E$ with $[P,Q] = E$ and all other brackets zero. Let $f = E^*$. Since $\mathfrak{g}(f) = \mathbb{R}E$, any real polarization must have dimension 2 and contain E. For example $\mathfrak{h} = \mathbb{R}P+\mathbb{R}E$, or $\mathfrak{h} = \mathbb{R}Q+\mathbb{R}E$.

Another fact is that a maximal totally isotropic subspace for B_f need not be a real polarization (because it may fail to be a subalgebra).

EXAMPLE. Let $\mathfrak{g}$ be a four-dimensional Lie algebra with generators x_i, $1 \leq i \leq 4$, satisfying $[x_1,x_2] = x_3$, $[x_1,x_3] = x_4$. If $f = x_4^*$, then $\mathfrak{g}(f) = \mathbb{R}x_2 + \mathbb{R}x_4$. The space $\mathfrak{h} = \mathbb{R}x_1 + \mathbb{R}x_2 + \mathbb{R}x_4$ is a maximal totally isotropic subspace, although it is not an algebra.

Now let G be a connected and simply connected nilpotent Lie group, $\mathfrak{g} = LA(G)$, $f \in \mathfrak{g}^*$. Let $\mathfrak{h}$ be a subalgebra such that $f([\mathfrak{h},\mathfrak{h}]) = 0$, and let H be the connected (and hence simply connected) subgroup corresponding to $\mathfrak{h}$. Then $H = \exp \mathfrak{h}$ and we can define a character of H by

$$\chi_f(\exp x) = e^{if(x)}, \qquad x \in \mathfrak{h}.$$

Let $\rho(f,\mathfrak{h},G) = \mathrm{Ind}_H^G \chi_f$.

THEOREM 1. *Let* G *be a simply connected nilpotent Lie group,* $\mathfrak{g} = LA(G)$, $f \in \mathfrak{g}^*$.

(i) *There exists a real polarization at* f.

(ii) *For* $\mathfrak{h}$ *a real polarization at* f, *the representation* $\rho(f,\mathfrak{h},G)$ *is irreducible.*

(iii) *If* $\mathfrak{h}_1$, $\mathfrak{h}_2$ *are two real polarizations at* f, *then* $\rho(f,\mathfrak{h}_1,G) \cong \rho(f,\mathfrak{h}_2,G)$.

The proof is a lengthy induction on $\dim \mathfrak{g}$. First, if $\dim \mathfrak{g} = 1$, then $G = \mathbb{R}$ and for any $f \in \mathfrak{g}^*$, $\mathfrak{h} = \mathfrak{g}$ and we get the characters of G. Let $\dim \mathfrak{g} > 1$ and assume the result for all nilpotent groups of lower dimension. Let $\mathfrak{z} = \mathrm{cent}\, \mathfrak{g}$. Then $\ker f \cap \mathfrak{z} = I$ is an abelian ideal of codimension one in $\mathfrak{z}$.

Case 1. $I \neq \{0\}$. I is of course an ideal in $\mathfrak{g}$. Consider the quotient $\mathfrak{g}' = \mathfrak{g}/I$ and write $\pi\colon \mathfrak{g} \to \mathfrak{g}'$ for the canonical projection. Choose $f' \in \mathfrak{g}'^*$ such that $f' \circ \pi = f$.

(i) By induction, there is $\mathfrak{h}' \subseteq \mathfrak{g}'$ satisfying (i), i.e.

$f'([\mathfrak{h}', \mathfrak{h}']) = 0$ and $f'([x, \mathfrak{h}']) \neq 0$ if $x \notin \mathfrak{h}'$. Then an easy computation shows that $\mathfrak{h} = \pi^{-1}\mathfrak{h}'$ satisfies (i) for f.

(ii) Let $\mathfrak{h}$ be as in (i). Then $\mathfrak{h} \supseteq I$ because $x \in I \Rightarrow [x,I] \subseteq I \Rightarrow f([x,I]) \in f(I) = 0$. Lift the projection to the corresponding groups, $\pi: G \to G'$. Then clearly $\rho(f,\mathfrak{h},G) = \rho(f',\mathfrak{h}',G') \circ \pi$. Since $\rho(f',\mathfrak{h}',G')$ is irreducible, it follows that $\rho(f,\mathfrak{h},G)$ is also irreducible.

(iii) Apply the same technique as in (ii). Let $\mathfrak{h}_1' = \mathfrak{h}_1/I$, $\mathfrak{h}_2' = \mathfrak{h}_2/I$. Then the equivalence of $\rho(f',\mathfrak{h}_1',G')$ with $\rho(f',\mathfrak{h}_2',G')$ lifts naturally to an equivalence of $\rho(f,\mathfrak{h}_1,G)$ with $\rho(f,\mathfrak{h}_2,G)$.

Case 2. $I = \{0\}$. This forces $\mathfrak{z}$ to be of dimension 1 and $f(\mathfrak{z}) \neq \{0\}$. We may choose $Z \in \mathfrak{z}$ such that $f(Z) = 1$. Now since $\mathfrak{g}$ is nilpotent and $\mathfrak{z} \neq \mathfrak{g}$, we have $C_2(\mathfrak{z}) = \{x: [x,\mathfrak{g}] \subseteq \mathfrak{z}\} \supsetneq \mathfrak{z}$. Choose $Y \in C_2(\mathfrak{z})-\mathfrak{z}$. Replacing Y by $Y-f(Y)Z$ if necessary, we may assume $f(Y) = 0$. Next set $\mathfrak{a} = \mathbb{R}Y + \mathbb{R}Z$. Then $\mathfrak{a}$ is an abelian subalgebra of $\mathfrak{g}$, in fact it's an ideal since $[\mathfrak{g},Y] \subseteq \mathfrak{z}$.

Next consider the ideal $\mathfrak{g}'$ of $\mathfrak{g}$ given by

$$\mathfrak{g}' = \{x \in \mathfrak{g}: [x,\mathfrak{a}] = 0\} = \{x \in \mathfrak{g}: [x,Y] = 0\}.$$

There is a linear form λ such that $[x,Y] = \lambda(x)Z$, therefore $\mathfrak{g}' = \ker \lambda$ must be of codimension 1 in $\mathfrak{g}$.

Let $f' = f|_{\mathfrak{g}'}$. We are going to apply the induction hypothesis to $\mathfrak{g}'$.

(i) Let $\mathfrak{h}'$ be a subalgebra of $\mathfrak{g}'$ such that $f'([x,\mathfrak{h}']) = 0 \Leftrightarrow x \in \mathfrak{h}'$. Then $\mathfrak{a} \subseteq \mathfrak{h}'$ because $[\mathfrak{h}',\mathfrak{a}] \subseteq [\mathfrak{g}',\mathfrak{a}] = \{0\}$. Let $\mathfrak{h} = \mathfrak{h}'$. Then $\mathfrak{h}$ satisfies (i) in $\mathfrak{g}$. In fact, $\mathfrak{h}$ is clearly isotropic -- we need only show maximality. Let $x \in \mathfrak{g}$ be such that $f([x,\mathfrak{h}]) = 0$. Then $f([x,Y]) = 0$ since $Y \in \mathfrak{h}$. But $[x,Y] = \lambda(x)Z$ and $f(Z) = 1 \Rightarrow \lambda(x) = 0$, that is $x \in \mathfrak{g}'$. By maximality in $\mathfrak{g}'$, we are done.

(ii) Let $\mathfrak{h}$ be a subalgebra of $\mathfrak{g}$ satisfying (i). We show $\rho(f,\mathfrak{h},G)$ is irreducible. Suppose first that $\mathfrak{h} \subseteq \mathfrak{g}'$. Then $\rho(f,\mathfrak{h},G) = \mathrm{Ind}_H^G \chi_f = \mathrm{Ind}_{G'}^G \mathrm{Ind}_H^{G'} \chi_f = \mathrm{Ind}_{G'}^G \rho(f',\mathfrak{h}',G')$, where we have written $G' = \exp \mathfrak{g}'$, $\mathfrak{h} = \mathfrak{h}'$ in $\mathfrak{g}'$. We now apply the group extension procedure of Chapter III to the normal subgroup $\exp \mathfrak{a}$ in G.

LEMMA 2. *Let* $A = \exp \mathfrak{a}$, *an abelian normal subgroup of* G. *Let* $\lambda_f(\exp x) = e^{if(x)}$, $x \in \mathfrak{a}$, *a unitary character of* A. *Then the stabilizer of* λ_f *in* G *is precisely* G'. *Moreover the restriction of* $\rho(f',\mathfrak{h}',G')$ *to* A *is a multiple of* λ_f.

Suppose the lemma proven. Since $\rho(f',\mathfrak{h}',G')$ is irreducible (the induction hypothesis), an immediate application of Chapter III, Theorem B2 gives the irreducibility of $\rho(f,\mathfrak{h},G)$.

Proof of Lemma 2. Let G_1 be the stability group of λ_f in G; $G_1 = \{g \in G: \lambda_f(gag^{-1}) = \lambda_f(a),\ \text{all } a \in A\}$. A simple calculation shows then that $G_1 = \{g \in G: \mathrm{Ad}(g)x - x \in \ker f,\ \text{for all } x \in \mathfrak{a}\}$. Writing $g = \exp y$ and using the fact that $\mathfrak{a} \subseteq C_2(\mathfrak{z})$, we find next that $G_1 = \{\exp y: [y,\mathfrak{a}] \subseteq \ker f\} = \{\exp y: [y,\mathfrak{a}] = 0\} = \exp \mathfrak{g}' = G'$.

Now let $x \in \mathfrak{a} = \mathbb{R}Y+\mathbb{R}Z$. The representation $\rho(f',\mathfrak{h}',G')$ is realized by left translations in the space of functions ϕ satisfying: $\phi(g'h') = \chi_f(h')^{-1}\phi(g')$; $\int_{G'/H'} |\phi|^2 < \infty$. Restricting to A, we compute for $a = \exp x$

$$\begin{aligned}\rho(f',\mathfrak{h}',G')(a)\phi(g') &= \phi(a^{-1}g') = \phi(\exp(-x)g') \\ &= \phi(g' \exp(-x)) = \chi_f(\exp x)\phi(g') \\ &= e^{if(x)}\,\phi(g').\end{aligned}$$

That is $\rho(f',\mathfrak{h}',G')|_A = \lambda_f I$. This completes the proof of Lemma 2.

However we still have to handle the possibility that $\mathfrak{h} \not\subseteq \mathfrak{g}'$. Then we can choose $X \in \mathfrak{h}$, $\mathfrak{g} = \mathfrak{g}' \oplus \mathbb{R}X$, $[X,Y] = Z$ and $f(X) = 0$ (replacing X by $X-f(X)Z$ if necessary). Then put $\mathfrak{h}_0 = \mathfrak{h} \cap \mathfrak{g}'$, $\mathfrak{h}' = \mathfrak{h}_0 + \mathfrak{a}$ and $\mathfrak{k} = \mathfrak{h} + \mathfrak{a}$. For example if $\mathfrak{g} = \mathfrak{h}_1$, $\mathfrak{a} = \mathbb{R}Q+\mathbb{R}E$, $\mathfrak{g}' = \mathfrak{a}$, $\mathfrak{h} = \mathbb{R}P+\mathbb{R}E$, $\mathfrak{h}_0 = \mathbb{R}E$, $\mathfrak{h}' = \mathbb{R}E+\mathbb{R}Q$ and $\mathfrak{k} = \mathfrak{g}$.

LEMMA 3. (i) $\mathfrak{h} = \mathfrak{h}_0 \oplus \mathbb{R}X$.

(ii) $\mathfrak{h}' = \mathfrak{h}_0 \oplus \mathbb{R}Y$ *and* $\mathfrak{h}'$ *is a real polarization at* f.

(iii) $\mathfrak{k} = \mathfrak{h}_0 \oplus \mathbb{R}X \oplus \mathbb{R}Y$ *is a subalgebra and every element in* $K = \exp \mathfrak{k}$ *can be written in a unique way as* $\exp uX \cdot h_0 \cdot \exp vY$, $h_0 \in \exp \mathfrak{h}_0 = H_0$, $u,v \in \mathbb{R}$.

Proof. (i) $\mathfrak{h}_0$ is of codimension 1 in $\mathfrak{h}$ and $X \notin \mathfrak{g}'$. The result follows by dimension.

(ii) First $f([\mathfrak{h}', \mathfrak{h}']) = f([\mathfrak{h}_0, \mathfrak{h}_0])$ since $[\mathfrak{g}', \mathfrak{a}] = [\mathfrak{a}, \mathfrak{a}] = 0$. Hence $\mathfrak{h}'$ is totally isotropic for B_f (since $\mathfrak{h}$ is). But $Y \notin \mathfrak{h}_0$ because $Y \notin \mathfrak{h}$. That is because $f[X,Y] = f(Z) = 1$. Hence $\mathfrak{h}' \supseteq \mathfrak{h}_0 + \mathbb{R}Y \supsetneq \mathfrak{h}_0$. But $\dim(\mathfrak{h}_0+\mathbb{R}Y) = \dim \mathfrak{h} \Rightarrow \dim(\mathfrak{h}_0+\mathbb{R}Y) = \dim \mathfrak{h}' \Rightarrow \mathfrak{h}' = \mathfrak{h}_0+\mathbb{R}Y$. That proves (ii).

(iii) Now $\mathfrak{k} = \mathfrak{h}_0+\mathbb{R}X+\mathbb{R}Y$ is an algebra because: $[X,Y] = Z \in \mathfrak{h}_0$, $[Y,\mathfrak{h}_0] \subseteq [Y,\mathfrak{g}'] \subseteq [\mathfrak{a}, \mathfrak{g}'] = \{0\}$, and $[X,\mathfrak{h}_0] \subseteq \mathfrak{h} \cap \mathfrak{g}'$ since $[[X,\mathfrak{h}_0],\mathfrak{a}] \subseteq [[\mathfrak{h}_0, \mathfrak{a}],X] + [[\mathfrak{a},X],\mathfrak{h}_0] \subseteq [[\mathfrak{g}',\mathfrak{a}],X] + [Z,\mathfrak{h}_0] = \{0\}$. We have already seen that $Y \notin \mathfrak{h}_0$. For the same reason $X \notin \mathfrak{h}_0$. So the sum is direct and the group decomposition is therefore unique.

LEMMA 4. $\rho(f,\mathfrak{h},G) \cong \rho(f,\mathfrak{h}',G)$.

Note that once Lemma 4 is proven, we are reduced to the previous case $\mathfrak{h} \subseteq \mathfrak{g}'$ which is already handled. Hence proving Lemma 4 completes the proof of (ii).

Proof of Lemma 4. By induction in stages, it is enough to prove

$$\rho(f, \mathfrak{h}, K) = \operatorname{Ind}_H^K \chi_f \cong \operatorname{Ind}_{H'}^K \chi_f' = \rho(f, \mathfrak{h}', K).$$

In some sense, this observation reduces matters to the case of a Heisenberg group.

Let $\mathcal{H}(f, \mathfrak{h}, K)$ be the space of the representation $\rho(f, \mathfrak{h}, K)$,

$$\mathcal{H}(f, \mathfrak{h}, K) = \{\phi\colon K \to \mathbb{C},\quad \phi(gh) = \chi_f(h)^{-1}\phi(g),\quad \int_{K/H} |\phi|^2 < \infty\}.$$

Similarly

$$\mathcal{H}(f, \mathfrak{h}', K) = \{\phi\colon K \to \mathbb{C},\quad \phi(gh') = \chi_f(h')^{-1}\phi(g),\quad \int_{K/H'} |\phi|^2 < \infty\}.$$

Both induced actions are by left translations. So we want to find a unitary operator $T_{\mathfrak{h}, \mathfrak{h}'}\colon \mathcal{H}(f, \mathfrak{h}, K) \to \mathcal{H}(f, \mathfrak{h}', K)$ that commutes with left translation by elements of K. We first give the general idea, then describe the actual construction.

Given $\phi \in \mathcal{H}(f, \mathfrak{h}, K)$, we have $\phi(gh) = \chi_f(h)^{-1}\phi(g)$. We want $(T_{\mathfrak{h}, \mathfrak{h}'}\phi)(gh') = \chi_f(h')^{-1}(T_{\mathfrak{h}, \mathfrak{h}'}\phi)(g)$. For $h \in H_0 = H \cap H' = \exp(\mathfrak{h} \cap \mathfrak{h}')$ that works automatically. It remains to handle $\exp \mathbb{R}Y$. Suppose we put

$$T_{\mathfrak{h}, \mathfrak{h}'}\,\phi(g) = \int \phi(g \exp vY)dv.$$

This certainly commutes with left translations and satisfies the correct equivariance condition. The problem is to guarantee convergence of the integral. This is handled by identifying the space with $L_2(\mathbb{R})$ and appealing to ordinary Fourier analysis.

Let $k \in K$, $k = \exp xY\cdot h$, $h \in H$, $x \in \mathbb{R}$

$= \exp yX\cdot h'$, $h' \in H'$, $y \in \mathbb{R}$.

Then $\phi \in \mathcal{H}(f, \mathfrak{h}, K) \Rightarrow \phi(k) = \phi(\exp xY\cdot h) = \chi_f(h)^{-1}\phi(\exp xY)$, so ϕ is uniquely determined by its values on $\exp \mathbb{R}Y$. Let $\mathcal{D}$ be the dense subspace of $\mathcal{H}(f, \mathfrak{h}, K)$ consisting of all continuous functions compactly supported modulo H. Define $R\colon \mathcal{D} \to C_0(\mathbb{R})$ by $(\mathbb{R}\phi)(x) = \phi(\exp xY)$. This is clearly an isometry onto a dense subspace of

$L_2(\mathbb{R})$, and so extends uniquely to an isometry of $\mathcal{H}(f,\mathfrak{h},K)$ onto $L_2(\mathbb{R})$. Similarly we define $R': \mathcal{D}' \to C_0(\mathbb{R})$, $(R'\phi)(y) = \phi(\exp yX)$ which extends to an isometry $R': \mathcal{H}(f,\mathfrak{h}',K) \to L_2(\mathbb{R})$. Thus we have

$$\begin{array}{ccc} \mathcal{H}(f,\mathfrak{h},K) & \xrightarrow{T_{\mathfrak{h},\mathfrak{h}'}} & \mathcal{H}(f,\mathfrak{h}',K) \\ \downarrow R & & \downarrow R' \\ L_2(\mathbb{R}) & \xrightarrow{\tilde{T}_{\mathfrak{h},\mathfrak{h}'}} & L_2(\mathbb{R}) \end{array}$$

and we are reduced to computing $\tilde{T}_{\mathfrak{h},\mathfrak{h}'}$.

For continuous functions of compact support, write

$$\begin{aligned} \tilde{T}_{\mathfrak{h},\mathfrak{h}'}(R\phi)(y) &= T_{\mathfrak{h},\mathfrak{h}'}(\phi)(\exp yX) \\ &= \int \phi(\exp yX \exp vY)dv \\ &= \int \phi(\exp vY \exp -vY \exp yX \exp vY)dv \\ &= \int \phi(\exp vY \exp[Ad(\exp -vY)(yX)]dv \\ &= \int \phi(\exp vY \exp[e^{ad(-vY)}(yX)]dv \\ &= \int \phi(\exp vY \exp[yX + vyZ]dv \\ &= \int \phi(\exp vY \exp yX \exp vyZ)dv. \end{aligned}$$

Now $X \in \mathfrak{h}$, $Z \in \mathfrak{h}$ and $\phi \in \mathcal{H}(f,\mathfrak{h},K) \Rightarrow$

$$\begin{aligned} \phi(\exp vY \exp yX \exp vyZ) &= e^{-ivyf(Z)}\, e^{-iyf(X)}\, \phi(\exp vY) \\ &= e^{-ivy}\, \phi(\exp vY). \end{aligned}$$

Thus

$$\begin{aligned} \tilde{T}_{\mathfrak{h},\mathfrak{h}'}(R\phi)(y) &= \int e^{-ivy}\, \phi(\exp vY)dv \\ &= \int (R\phi)(v)\, e^{-ivy}\, dv. \end{aligned}$$

That is the operator $\tilde{T}_{\mathfrak{h},\mathfrak{h}'}$ is nothing more than the Fourier transform. Therefore it extends to a unitary operator $\mathcal{F}$ on $L_2(\mathbb{R})$ by the classical Plancherel Theorem. Thus $T_{\mathfrak{h},\mathfrak{h}'} = R'^{-1}\mathcal{F}R$.

(iii) This is done in exactly the same manner as in (ii). If

$\mathfrak{h}_1$ and $\mathfrak{h}_2$ are both in $\mathfrak{g}'$, we proceed by the Mackey theory argument and the induction hypothesis. If not, we find $\mathfrak{h}_1'$ and/or $\mathfrak{h}_2'$ in $\mathfrak{g}'$ and apply the reasoning of (ii). That completes the proof of (iii) and Theorem 1.

We now indicate the proof of an equally important result due to Kirillov [1] and Dixmier [3] (see also Takenouchi [1]).

THEOREM 5. *Every irreducible representation of* G *is monomial, that is induced by a one-dimensional representation of some (connected) subgroup.*

Proof. (Sketch) The proof proceeds by induction on dim G. Let $\pi \in \hat{G}$, $Z = \text{Cent } G$, $\dim Z \geq 1$. There is a character $\chi \in \hat{Z}$ such that $\pi(z) = \chi(z)I$, $z \in Z$. Let $Z_0 = \ker \chi$. If $\dim Z_0 > 0$, then π is trivial on the neutral component H of Z_0. Hence π is lifted from an irreducible representation $\tilde{\pi}$ of G/H. By the induction hypothesis $\tilde{\pi} = \text{Ind}_{K_1}^{G/H} \chi_1$, χ_1 a character of $K_1 \subseteq G/H$. Then $\pi = \text{Ind}_K^G \chi$, where K is the inverse image of K_1 and χ is the lift of χ_1 to K.

Thus we may assume $\dim Z = 1$ and χ is not trivial.

LEMMA 5a. (Kirillov [1]) *Any such representation* π *is induced from a (necessarily irreducible) representation of a subgroup of codimension* 1.

Proof. (Sketch) The proof is by showing that G contains a 3-dimensional normal Heisenberg group N with Z as center and a compatible codimension 1 subgroup G_0. $\pi|_N$ is then known explicitly, and one can compute the action of G_0 on N so as to be able eventually to employ the imprimitivity theorem. We refer to Kirillov [1, §4] for the details.

Finally using the induction hypothesis we have

$$\pi = \mathrm{Ind}_{G_0}^{G}\, \sigma \;=\; \mathrm{Ind}_{G_0}^{G}\; \mathrm{Ind}_{H}^{G}\, \chi = \mathrm{Ind}_{H}^{G}\, \chi.$$

Now from theorem 1 we know there is a mapping

$$\rho : \mathfrak{g}^* \to \hat{G}$$

obtained by putting $\rho(f) = \rho(f,\mathfrak{h},G)$ where $\mathfrak{h}$ is *any* real polarization at f. By Theorem 5 this mapping is surjective. Now G acts on $\mathfrak{g}^*$ by the co-adjoint representation $(g{\cdot}f)(x) = f(\mathrm{Ad}\, g^{-1}(x))$, $g \in G$, $x \in \mathfrak{g}$, $f \in \mathfrak{g}^*$. Write as usual $\mathfrak{g}^*/G$ for the orbit space. The final piece in the picture is

THEOREM 6. (Kirillov [1]) *The representations* $\rho(f_1)$ *and* $\rho(f_2)$, $f_1,f_2 \in \mathfrak{g}^*$, *are unitarily equivalent if and only if there exists* $g \in G$ *such that* $g{\cdot}f_1 = f_2$. *Hence there is a canonical bijection* $\mathfrak{g}^*/G \to \hat{G}$, $G{\cdot}f \to \rho(f)$.

Proof. (Sketch) As usual, we employ induction on $\dim G$. By the usual kind of reasoning, we reduce matters to the case $\dim Z = 1$ and the representations involved are non-trivial on Z. Then the group G_0 of codimension 1 comes into play. Let $f_i \in \mathfrak{g}^*$, $f_i^0 = f_i|_{\mathfrak{g}_0}$, $i = 1,2$. The key step in the argument is to show that

$$\rho(f_i,G) \cong \mathrm{Ind}_{G_0}^{G}\, \rho\,(f_i^0,G_0)$$

(see Kirillov [1]). Then suppose f_1 and f_2 are G-conjugate. One shows that in fact they must be G_0-conjugate $\Rightarrow f_1^0$ and f_2^0 are G_0-conjugate. By induction $\rho(f_1^0,G_0) \cong \rho(f_2^0,G_0)$. Therefore $\rho(f_1,G) \cong \rho(f_2,G)$. Conversely suppose $\rho(f_1,G) \cong \rho(f_2,G)$. The stabilizer of both representations $\rho(f_i^0,G_0)$, $i = 1,2$, is the group G_0 itself. Hence an application of the Mackey theory shows that $\rho(f_1^0,G_0)$ and $\rho(f_2^0,G_0)$ are G-conjugate. One then shows that they are in fact G_0-conjugate. Hence (by the induction hypothesis) f_1^0

and f_2^0 are G_0-conjugate. Finally one extends to get that f_1 and f_2 are G-conjugate.

This completes the arguments that establish the Kirillov correspondence $\mathfrak{g}^*/G \to \hat{G}$. We postpone consideration of examples until we are done with the remaining elements of the nilpotent theory we want to discuss. The first of these is the result (mentioned earlier) on the dimension of real polarizations. The following proof is taken from Quint [1].

THEOREM 7. *Let* $f \in \mathfrak{g}^*$ *with* $\mathfrak{h} \subseteq \mathfrak{g}$ *a totally isotropic subspace. Set* $\mathfrak{g}(f) = \{x \in \mathfrak{g}: f([x,\mathfrak{g}]) = 0\} = \{x: B_f(x,y) = 0 \ \forall y \in \mathfrak{g}\}$. *Then* $\mathfrak{h}$ *is maximal totally isotropic if and only if* $\dim \mathfrak{h} = \frac{1}{2}(\dim \mathfrak{g} + \dim \mathfrak{g}(f))$.

Proof. Let $\mathfrak{h}$ be a subspace of $\mathfrak{g}$. Set $\mathfrak{h}^f = \{x \in \mathfrak{g}: B_f(x,\mathfrak{h}) = 0\}$. Then $\mathfrak{h}$ is totally isotropic if and only if $\mathfrak{h}^f \supseteq \mathfrak{h}$.

(1) $\dim \mathfrak{h}^f = \dim \mathfrak{g} - \dim \mathfrak{h} + \dim(\mathfrak{h} \cap \mathfrak{g}(f))$.

Consider $\tilde{\mathfrak{g}} = \mathfrak{g}/\mathfrak{g}(f)$. If $\tilde{\mathfrak{h}}$ denotes the image of $\mathfrak{h}$ under the canonical projection $\mathfrak{g} \to \tilde{\mathfrak{g}}$, then $\dim \tilde{\mathfrak{g}} = \dim \tilde{\mathfrak{h}} + \dim \tilde{\mathfrak{h}}^f$. This is because on $\tilde{\mathfrak{g}}$, the form B_f is non-degenerate and $\tilde{\mathfrak{h}}^f$ is the orthogonal complement of $\tilde{\mathfrak{h}}$. Also $\dim \tilde{\mathfrak{h}} = \dim(\mathfrak{h}/\mathfrak{h} \cap \mathfrak{g}(f)) \Rightarrow \dim \mathfrak{h} = \dim \tilde{\mathfrak{h}} + \dim(\mathfrak{h} \cap \mathfrak{g}(f))$. Then we compute

$$\begin{aligned}
\dim \mathfrak{h}^f &= \dim \tilde{\mathfrak{h}}^f + \dim(\mathfrak{h}^f \cap \mathfrak{g}(f)) \\
&= \dim \tilde{\mathfrak{h}}^f + \dim \mathfrak{g}(f) \\
&= \dim \tilde{\mathfrak{g}} - \dim \tilde{\mathfrak{h}} + \dim \mathfrak{g}(f) \\
&= \dim \tilde{\mathfrak{g}} - \dim \mathfrak{h} + \dim(\mathfrak{g}(f) \cap \mathfrak{h}) + \dim \mathfrak{g}(f) \\
&= \dim \mathfrak{g} - \dim \mathfrak{h} + \dim(\mathfrak{g}(f) \cap \mathfrak{h}).
\end{aligned}$$

(2) $\mathfrak{h}$ is maximal totally isotropic $\Leftrightarrow \mathfrak{h} = \mathfrak{h}^f$.

In fact if $\mathfrak{h}' \supseteq \mathfrak{h}$ is totally isotropic, then $B_f(\mathfrak{h}', \mathfrak{h}') = 0$. Therefore $\mathfrak{h}' \subseteq \mathfrak{h}^f = \mathfrak{h} \Rightarrow \mathfrak{h}$ is maximal totally isotropic.

Conversely let $\mathfrak{h}$ be maximal totally isotropic. If $x \in \mathfrak{h}^f$, then $B_f(x,\mathfrak{h}) = 0$. Since $B_f(x,x) = 0$, we get that $\mathbb{R}x+\mathfrak{h}$ is totally isotropic. Therefore $x \in \mathfrak{h}$.

Finally if $\mathfrak{h}$ is totally isotropic, we show that

(3) $\mathfrak{h}$ is maximal $\iff \dim \mathfrak{h} = \frac{1}{2}(\dim \mathfrak{g} + \dim \mathfrak{g}(f))$.

Suppose first that $\mathfrak{h}$ is maximal. By (2), $\mathfrak{h} = \mathfrak{h}^f$. Then by (1), $\dim \mathfrak{h} = \dim \mathfrak{g} - \dim \mathfrak{h} + \dim(\mathfrak{h} \cap \mathfrak{g}(f))$. Hence $2 \dim \mathfrak{h} = \dim \mathfrak{g} + \dim \mathfrak{g}(f)$, since $\mathfrak{h} = \mathfrak{h}^f \supseteq \mathfrak{g}(f)$. Conversely, suppose $\dim \mathfrak{h} = \frac{1}{2}(\dim \mathfrak{g} + \dim \mathfrak{g}(f))$. Then by (1)

$$\begin{aligned}\dim \mathfrak{h}^f &= \frac{1}{2} \dim \mathfrak{g} - \frac{1}{2} \dim \mathfrak{g}(f) + \dim(\mathfrak{h} \cap \mathfrak{g}(f)) \\ &\leqq \frac{1}{2} \dim \mathfrak{g} + \frac{1}{2} \dim \mathfrak{g}(f) \\ &= \dim \mathfrak{h}.\end{aligned}$$

Since $\mathfrak{h} \subseteq \mathfrak{h}^f$, we conclude $\mathfrak{h} = \mathfrak{h}^f$. Therefore by (2), $\mathfrak{h}$ is maximal.

Let $\mathcal{O}_f$ denote the orbit of $f \in \mathfrak{g}^*$ under G. Then $\dim G - \dim G(f) = \dim \mathcal{O}_f = \dim \mathfrak{g} - \dim \mathfrak{g}(f)$.

COROLLARY. *Let* $\mathfrak{h}$ *be a real polarization of* f. *Then*

$$\dim \mathfrak{h} = \dim \mathfrak{g} - \frac{1}{2} \dim \mathcal{O}_f .$$

EXERCISE. Show that another consequence is that $\mathfrak{h} \subseteq \mathfrak{g}$ ($\mathfrak{g}$ nilpotent) is a real polarization at $f \iff \mathfrak{h}$ is a subalgebra of maximal dimension which is subordinate to f.

Note that the representation $\rho(f,\mathfrak{h},G)$ is defined for any subalgebra $\mathfrak{h}$ subordinate to f, maximal or otherwise. Kirillov [1] proved not only that $\mathfrak{h}$ maximal implies $\rho(f,\mathfrak{h},G)$ is irreducible, but the converse as well.

B. CHARACTERS AND THE PLANCHEREL FORMULA

Next we give a brief indication of the existence of characters and the form of the Plancherel measure μ_G on G. (Refer back to Chapter I section C for generalities on characters and μ_G.) This material is found in Kirillov [1, Chapter 7] which we summarize here.

We begin with a theorem whose proof involves an induction argument that essentially reduces to the Heisenberg group case.

THEOREM 1. *Every* $\pi \in \hat{G}$ *may be realized in* $L_2(\mathbb{R}^m)$ *in such a way that* $\pi(\mathfrak{U}(\mathfrak{g})) = \mathcal{D}_m$ = *the algebra of differential operators with polynomial coefficients. Furthermore* $m = \dim \mathfrak{g} - \dim \mathfrak{h} = \frac{1}{2} \dim \mathcal{O}_f$ *if* $\pi = \rho(f, \mathfrak{h}, G)$.

We shall see in a moment that a consequence of this result is that G is traceable. But first, we mention a corollary relating to the infinitesimal characters of G.

COROLLARY. *Let* $\pi = \rho(f, \mathfrak{h}, G)$, $\chi_\pi : \mathfrak{z} \to \mathbb{C}$ *the infinitesimal character of* π. *Then*

$$\chi_\pi(z) = z(f), \quad z \in \mathfrak{z}$$

in the following sense -- each $z \in \mathfrak{z}$ *is identified to a* G-*invariant polynomial function on* $\mathfrak{g}^*$ *and* $z(f)$ *is its value at the point* $f \in \mathfrak{g}^*$.

REMARK. In the nilpotent case, unlike the general case, we have the additional fact that the infinitesimal characters determine the irreducible representations uniquely. That this is so (at least for the representations in general position, i.e. those corresponding to orbits of maximal dimension) is seen immediately from the following result of Chevalley (see e.g. Pukanszky [3]).

THEOREM 2. *Let* V *be a finite-dimensional real vector space with* G *acting on* V *by unipotent matrices. Then there exist* G-*invariant polynomial functions* $p_0, p_1, \cdots, p_k$ *on* V *such that* $V_0 = \{v \in V: p_0(v) \neq 0\}$ *is fibered by* G *into* (n-k)-*dimensional orbits, given by the equations* p_i = *constant,* $1 \leq i \leq k$.

We have perhaps put the cart before the horse in computing the infinitesimal characters before showing that the irreducible representations have global characters. In fact as mentioned earlier, we have

THEOREM 3. *Connected, simply connected nilpotent Lie groups are traceable.*

Proof. Let $\pi = \rho(f, \mathfrak{h}, G)$ and let $m = \frac{1}{2} \dim \mathcal{O}_f$. Let M be any differentiable operator on $\mathbb{R}^m$ with polynomial coefficients whose inverse is a trace class operator. By Theorem 1, there is $p \in \mathfrak{U}(\mathfrak{g})$ such that $\pi(p) = M$. Let D be the differential operator on G corresponding to p. Then for $\phi \in C_0^\infty(G)$, we have

$$\pi(\phi) = M^{-1}M\pi(\phi) = M^{-1}\pi(p)\pi(\phi) = M^{-1}\pi(D\phi).$$

But $\pi(D\phi)$ is a bounded operator, so $\pi(\phi)$ is trace class.

EXERCISE. Show that $\Theta_\pi: \phi \to \mathrm{Tr}\ \pi(\phi)$ is continuous.

COROLLARY. G *is* CCR.

It is possible to give a more explicit determination of $\Theta_\pi, \pi \in \hat{G}$. This is due to Kirillov and Pukanszky and goes as follows. The exponential map $\exp: \mathfrak{g} \to G$ is a diffeomorphism of manifolds. Also, there is a natural Fourier transform from $\mathfrak{g}$ to $\mathfrak{g}^*$, namely

$$\tilde{\psi}(f) = \int_{\mathfrak{g}} e^{if(x)} \psi(x)dx, \qquad \psi \in C_0^\infty(\mathfrak{g}).$$

THEOREM 4. (Kirillov [1]) *Let* $\pi = \rho(f, \mathfrak{h}, G)$. *Then there is a G-invariant measure* μ_π *on* $\mathcal{O}_f$ *such that*

$$\operatorname{Tr} \pi(f) = \int_{\mathcal{O}_f} \tilde{\phi}(h)\,d\mu_\pi(h)$$

where we write $\phi(x) = \phi(\exp x)$, $x \in \mathfrak{g}$.

Another way to say the same thing is the following. The distribution Θ_π is obtained by taking the measure μ_π on $\mathcal{O}_f$, extending it to a measure on $\mathfrak{g}^*$, taking its Fourier transform to get a distribution on $\mathfrak{g}$, and then lifting to G via the exponential map.

There are three additional facts: (1) Θ_π is tempered in the sense that it's tempered on $\mathfrak{g}$ (that is, it can be extended to the Schwartz space of rapidly decreasing C^∞ functions); (2) the measure μ_π is uniquely specified up to a constant since $\mathcal{O}_f = G/G(f)$ carries a unique (up to a scalar) G-invariant measure; and (3) one may compute the normalization of μ_π in terms of the symplectic geometry of the Kirillov picture. Indeed B_f is non-degenerate on $\mathfrak{g}/\mathfrak{g}(f)$. Since the tangent space to $\mathcal{O}_f$ at f is canonically identified to $\mathfrak{g}/\mathfrak{g}(f)$, it carries a non-degenerate skew two-form ω_f. Set $\nu^0 = \omega_f \wedge \cdots \wedge \omega_f$, $k = \frac{1}{2}$ m-times, to get a volume element on $\mathcal{O}_f$. Then Pukanszky [4] showed that the proper normalization is $(k!(4\pi)^k)^{-1}\nu^0$.

Finally we comment on the Plancherel measure. Since G is CCR, it is type I. In addition nilpotent groups are unimodular ($\det \operatorname{Ad}_G(x) = 1$). Hence we know there is a measure μ_G (unique up to a normalization of Haar measure) on $\hat{G}$ such that

$$\int_G |\phi(g)|^2 dg = \int_{\hat{G}} \|\pi(\phi)\|_2^2 \, d\mu_G(\pi), \qquad \phi \in L_1(G) \cap L_2(G).$$

We have already seen that to compute μ_G it suffices to establish the inversion formula

$$\psi(e) = \int_{\hat{G}} \operatorname{Tr} \psi(\pi)\,d\mu_G(\pi), \qquad \psi \in C_0^\infty(G).$$

But we already know that

$$\mathrm{Tr}\ \pi_f(\psi) = \int_{\mathcal{O}_f} \tilde{\psi}(h) d\mu_f(h), \qquad \pi_f = \rho(f, \mathfrak{h}, G).$$

Thus, by the ordinary Fourier inversion formula, and by disintegration of measures (Chapter III, Theorem B5), there is a pseudo-image $\overline{\mu}$ of Lebesgue measure on $\mathfrak{g}^*$ such that

$$\begin{aligned} \phi(e) &= \int_{\mathfrak{g}^*} \tilde{\phi}(h) dh \\ &= \int_{\mathfrak{g}^*/G} \int_{G/G(f)} \tilde{\phi}(g\cdot h) d\overline{g}\ d\overline{\mu}(\overline{h}) \\ &= \int_{\hat{G}} \int_{\mathcal{O}_f} \phi(h) d\mu_f(h) d\overline{\mu}(f) \\ &= \int_{\hat{G}} \mathrm{Tr}\ \pi_f(\phi)\ d\overline{\mu}(f). \end{aligned}$$

Hence the Plancherel measure on $\hat{G}$ is *the* pseudo-image of Lebesgue measure on $\mathfrak{g}^*/G$ with the normalizations of the measures on the orbits as chosen previously. In fact it is possible to be somewhat more precise about $\mu_G = \overline{\mu}$ as follows.

The set Λ of orbits in general position is Zariski open in $\mathfrak{g}^*$. In Λ there are natural coordinates, namely if $p_1, \cdots, p_k$ are the G-invariant polynomials of Theorem 2, then for the coordinates of an oribt $\mathcal{O}$, we take $\lambda_i = p_i(\mathcal{O})$, $1 \leq i \leq k$. The final result is

THEOREM 5. (Kirillov [1]) *There exists a rational function* $R(\lambda_1, \cdots, \lambda_k)$ *such that*

$$\phi(e) = \int_{\Lambda} \mathrm{Tr}\pi_\lambda(\phi)\ |R(\lambda)| d\lambda$$

where $\lambda = (\lambda_1, \cdots, \lambda_k)$ and $d\lambda = d\lambda_1 \cdots d\lambda_k$.

We now conclude this chapter with the promised examples.

EXAMPLES. (1) *Heisenberg groups.* For $n \geq 1$, set

$$G = \mathfrak{N}_n = \{\begin{pmatrix} 1 & x_1\cdots x_n & & z \\ & 1 & & y_1 \\ & & \ddots \; 0 & \vdots \\ & & \ddots & y_n \\ 0 & & & 1 \\ & & & 1 \end{pmatrix} : \text{ real entries}\}.$$

This is a two-step nilpotent group $G = (x_1,\cdots,x_n;\ y_1,\cdots,y_n;z)$. The commutator subgroup and the center are both $Z = \{(0;0;z) : z \in \mathbb{R}\}$ and $G/Z \cong \mathbb{R}^{2n}$. The Lie algebra $\mathfrak{h}_n$ is given by

$$\mathfrak{h}_n = \{\begin{pmatrix} 0 & x_1\cdots x_n & z \\ & \ddots & y_1 \\ & \ddots \; 0 & \vdots \\ & \ddots & y_n \\ 0 & & 0 \end{pmatrix} : \text{ real entries}\},$$

and so has generators $P_1,\cdots,P_n;\ Q_1,\cdots,Q_n;\ E$ with relations $[P_i,Q_i] = E$, $1 \leq i \leq n$, all other brackets zero. We denote the dual basis P_i^*,Q_i^*,E^*. It's a simple check that if $f \in \mathfrak{h}_n^*$ and $f(E) = 0$, then $\mathcal{O}_f = \{f\}$. Also any subalgebra is subordinate, and hence $\mathfrak{g}$ itself is a real polarization. The resulting representations are the characters

$$\pi_f(\exp(\sum\alpha_i P_i + \sum\beta_j Q_j + \gamma E)) = e^{if(\sum\alpha_i P_i + \sum\beta_j Q_j)}.$$

These are the representations of G, trivial on Z, i.e. $(G/Z)^{\wedge} \cong \mathbb{R}^{2n}$.

If $f(E) \neq 0$, then $\mathcal{O}_f = \{h \in \mathfrak{h}_n^* : h(Z) = f(Z)\}$, a 2n-dimensional hyperplane through f. We can parameterize the orbits in general position then by αE^*, $\alpha \neq 0$. A maximal subordinate subalgebra is obtained by setting $\mathfrak{h} = \mathbb{R}E + \sum_{i=1}^{n} \mathbb{R}P_i$. In fact any real polarization is of the form $\mathbb{R}E+W$, where W is a maximal totally isotropic subspace of $\sum\mathbb{R}P_i+\sum\mathbb{R}Q_j$ for B_f. Hence the representations $\rho(\alpha E^*,\mathfrak{h},G) = \pi_\alpha$ are infinite-dimensional representations obtained by exponentiating $\alpha E^*|_{\mathfrak{h}}$ to $\exp\mathfrak{h}$ and inducing to $\mathfrak{N}_n$.

The characters of this group are as follows: for the one-dimensional representations, they are the representations themselves; for the representations π_α the characters are distributions, supported on the one-dimensional subgroup Z and equal to the measure $c_\alpha e^{2\pi i\alpha z}dz$ there. As for the infinitesimal characters, the algebra $\mathfrak{z}(\mathfrak{g})$ is one-dimensional and

$$\chi_\pi(\zeta) = 2\pi i\alpha\zeta, \qquad \zeta \in \mathfrak{z}.$$

Finally for the Plancherel measure, the rational function of Theorem 5 is precisely α^n. Hence the Plancherel formula becomes

$$\int_{\mathcal{H}_n} |\phi(g)|^2 \, dg = c_n \int_{\hat{G}} \|\pi_\alpha(\phi)\|_2^2 \, |\alpha|^n \, d\alpha.$$

(2) Let G_n be the upper triangular $n\times n$ real unipotent matrices,

$$G_n = \{ \begin{pmatrix} 1 & & * \\ & \ddots & \\ 0 & & 1 \end{pmatrix} : \text{ real entries}\}.$$

$\mathfrak{g}_n = LA(G_n)$ is the upper triangular real nilpotent matrices. We identify $\mathfrak{g}_n^*$ with the space of lower triangular nilpotent matrices. The co-adjoint representation is then $g\cdot\phi = (g\phi g^{-1})_{low} = ((g\phi g^{-1})_{ij}\varepsilon_{ij})$ where $\varepsilon_{ij} = \{\begin{matrix} 1 & i > j \\ 0 & i \not> j \end{matrix}\}$.

Let Δ_k be the determinant of the left minor of order k, $k = 1,\cdots,[\frac{n}{2}]$. These polynomials form a basis in the algebra of all G_n-invariant polynomials in $\mathfrak{g}_n^*$. Thus an orbit in general position is given by the equations $\Delta_k(\phi) = \text{constant} \neq 0$. The dimension of such an orbit is $\frac{n(n-1)}{2} - [\frac{n}{2}]$. Thus a polarization must have dimension $[\frac{n}{2}] \times [\frac{n+1}{2}]$. In fact $\mathfrak{g}_n$ has an abelian ideal $\mathfrak{h}$ of this dimension, namely all matrices $\mathfrak{h} = \{\begin{pmatrix} 0 & p \\ 0 & 0 \end{pmatrix} : p$ an $[\frac{n}{2}] \times [\frac{n+1}{2}]$ matrix$\}$. $\mathfrak{h}$ is a real polarization. Then $G/H \cong \{\begin{pmatrix} g_1 & 0 \\ 0 & g_2 \end{pmatrix} : g_1 \in G_{[\frac{n}{2}]},$

$g_2 \in G_{[\frac{n+1}{2}]}\}$. As representatives for the orbits in general position, we take $\begin{pmatrix} 0 & 0 \\ \Lambda & 0 \end{pmatrix}$, where

$$\Lambda = \begin{pmatrix} 0 & & \lambda_k \\ & \lambda_2 \cdot^{\cdot^{\cdot}} & \\ \lambda_1 & & 0 \end{pmatrix}, \quad n = 2k \qquad\qquad \Lambda = \begin{pmatrix} 0 & \cdots & 0 \\ & \cdot^{\cdot^{\cdot}} & \lambda_k \\ \lambda_1 & & \\ & & 0 \end{pmatrix}, \quad n = 2k+1.$$

The resulting representations of G look like

$$\pi_g^{\Lambda} f(X_1,X_2) = e^{i\,tr(\Lambda X_1 B A_2^{-1} X_2^{-1})} f(X_1A_1, X_2A_2), \qquad f \in L_2(G/H)$$

if $X = \begin{pmatrix} X_1 & 0 \\ 0 & X_2 \end{pmatrix} \in G/H, \quad g = \begin{pmatrix} A_1 & B \\ 0 & A_2 \end{pmatrix}$.

Kirillov [1] writes down the character of these representations -- the formulas are quite complicated. The Plancherel formula is somewhat easier to describe. Indeed

$$\int_{G_n} |\phi(g)|^2 dg = \int_{\hat{G}_n} \|\pi^{\Lambda}(\phi)\|_2^2 \, d\mu(\Lambda)$$

where

$$d\mu(\Lambda) = \begin{cases} (2\pi)^{-k^2} \lambda_2^2\lambda_3^4\cdots\lambda_k^{2k-2} \, d\lambda_1\cdots d\lambda_k, & n = 2k \\ (2\pi)^{-k(k+1)} |\lambda_1\lambda_2^3\cdots\lambda_k^{2k-1}| \, d\lambda_1\cdots d\lambda_k, & n = 2k+1. \end{cases}$$

(3) Of course there is no complete classification theory for nilpotent groups as there is for semisimple groups. There have been some attempts at classification though for low-dimensional groups. For example, Dixmier [4] classified all G of dimension ≤ 5 and gave their representation theory. For the sake of another example and to fulfill a promise from Chapter I, p. 11, we give one of his examples.

Let $\mathfrak{g} = \sum_{i=1}^{4} \mathbb{R}X_i$ where the generators X_i satisfy $[X_1,X_2] = X_3$, $[X_1,X_3] = X_4$, all other brackets zero. Let $G = \exp \mathfrak{g}$. The following computations are tedious but straightforward:

$g = \exp X$, $X = \sum x_iX_i$, $Y = \sum y_iX_i$, $f(Y) = \sum f_iy_i$, $f_i \in R$, $f \in \mathfrak{g}^*$,

$$\mathrm{Ad}\, g(X) = y_1X_1+y_2X_2+(y_3+x_1y_2-x_2y_1)X_3+(y_4+x_1y_3-x_3y_1+\tfrac{1}{2}x_1^2y-\tfrac{1}{2}x_1x_2y_1)X_4$$

$$\mathrm{Ad}\, g^{-1}(f) = (f_1-x_2f_3-x_3f_4-\tfrac{1}{2}x_1x_2f_4, f_2+x_1f_3+\tfrac{1}{2}x_1^2f_4, f_3+x_1f_4, f_4).$$

There are three types of orbits:

(1) $f_3 = f_4 = 0$. These orbits are points and give rise to the two-parameter family of characters of G which are trivial on the normal subgroup $N_0 = \exp(\mathbb{R}X_3+\mathbb{R}X_4)$.

(2) $f_4 = 0$, $f_3 \neq 0$. The orbits are then $(f_1-x_2f_3, f_2+x_1f_3, f_3, 0)$, hyperplanes of dimension 2. The set $\{(0,0,f_3,0),\ f_3 \neq 0\}$ is a cross-section for these orbits. The stability subalgebra for $f = (0,0,f_3,0)$ is $\mathfrak{g}(f) = \mathbb{R}X_3+\mathbb{R}X_4$ and a polarization is $\mathbb{R}X_1+\mathbb{R}X_3+\mathbb{R}X_4$ or $\mathbb{R}X_2+\mathbb{R}X_3+\mathbb{R}X_4$. Thus we get a one-parameter family of representations induced from a subgroup of codimension 1. These representations are trivial on the normal subgroup $N_1 = \exp \mathbb{R}X_4$ -- they correspond to the infinite-dimensional representations of the Heisenberg group G/N_1.

(3) $f_4 \neq 0$. The orbits here are two-dimensional hypersurfaces and for a cross-section we may choose $\{(0,f_2,0,f_4),\ f_2 \in \mathbb{R},\ f_4 \neq 0\}$. The stability subalgebra for such an f is $\mathbb{R}X_2+\mathbb{R}X_4$ and for a polarization we must choose $\mathbb{R}X_2+\mathbb{R}X_3+\mathbb{R}X_4$. So we get a two-parameter family of representations, again induced from a subgroup of codimension 1.

Dixmier [4] has computed the Plancherel measure for the group. It is concentrated in the set of representations corresponding to the orbits (3), and is given by df_2df_4 there.

CHAPTER V. REPRESENTATIONS OF ALGEBRAIC GROUPS

Our main goals in this chapter are twofold. First, the groups we have been studying (semisimple Lie groups, nilpotent Lie groups) have analogs over local fields other than the real or complex numbers. The study of these is best couched in the framework of algebraic groups. Second we are interested in applying the Mackey procedure in order to study the representation theory of various kinds of mixtures of these groups. The beautiful structure theory of algebraic groups lends itself nicely to such a study.

A. STRUCTURE OF ALGEBRAIC GROUPS IN CHARACTERISTIC ZERO

We begin by recalling some facts about locally compact fields. In fact it is known that every locally compact non-discrete field is contained in the following list: (1) connected case -- $\mathbb{R}$ or $\mathbb{C}$; (2) totally disconnected case -- (a) characteristic zero - finite extensions of the p-adic numbers $\mathbb{Q}_p$; (b) characteristic $p > 0$ - the field of power series $\sum_{m}^{\infty} a_i t^i$, $m > -\infty$, where the coefficients a_i are from a finite field of characteristic p.

Let k be a non-discrete locally compact field. Until further notice we assume k is totally disconnected. We denote the Haar measure on k by dx. Then for $x_0 \in k$, there is a non-negative number $|x_0|$ such that $d(x_0 x) = |x_0| dx$. This defines a continuous function $x \to |x|$, $k \to [0,\infty)$ with the properties: $|x| = 0 \iff x = 0$; $|xy| = |x|\,|y|$, and $|x+y| \leq \max(|x|,|y|)$. The values of $|x|$ range over a set $\{q^n\colon n \in \mathbf{Z}\}$, q fixed. Hence the sets $\{x\colon |x| = c\}$, $\{x\colon |x| < c\}$ are open in k. In fact the latter constitute a neighborhood basis of 0.

THEOREM 1. (i) *The set* $\mathcal{O} = \{x \in k\colon |x| \leq 1\}$ *is compact and open in* k. *It is called the* ring of integers.

(ii) *The set* $\mathfrak{p} = \{x \in k\colon |x| < 1\}$ *is a prime ideal of* $\mathcal{O}$ *and the residue class field* $\mathcal{O}/\mathfrak{p}$ *has* $q = p^j$ *elements.*

(iii) $\mathfrak{p}$ *is a principal ideal, i.e.* $\mathfrak{p} = \wp\mathcal{O}$ *for some* $\wp$, *and* $|\wp| = q^{-1}$.

(iv) *There is* $\varepsilon \in k^*$ *of order* $q-1$ *such that* $|\varepsilon| = 1$ *and* $\{0, \varepsilon, \varepsilon^2, \cdots, \varepsilon^{q-1} = 1\}$ *forms a set of representatives for* $\mathcal{O}/\mathfrak{p}$.

(v) *Every non-zero* $x \in k$ *has a unique representation* $x = \wp^n(a_0 + a_1\wp + a_2\wp^2 + \cdots)$, $a_0 \neq 0$, $n \in \mathbb{Z}$, $a_i \in \{0, \varepsilon, \cdots, \varepsilon^{q-1}\}$.

EXAMPLE. If $k = \mathbb{Q}_p = \{\sum_{j > -\infty} b_j p^j$, then $\mathcal{O} = \{\sum_0^\infty b_j p^j\}$, $\mathfrak{p} = \{\sum_1^\infty b_j p^j\}$, $\mathcal{O}/\mathfrak{p}$ is generated by p and $|p| = p^{-1}$. Note that the series in (v) and the above series expansion of a p-adic number are not the same.

Next we briefly consider the characters of the abelian groups k and k^*. First, the group k is self-dual. Indeed if $\chi \in \hat{k}$ is any non-trivial character, then every character of k is of the form $\chi_u(x) = \chi(ux)$, $u \in K$. The map $u \to \chi_u$ is a topological isomorphism of k with $\hat{k}$. For k^*, matters are somewhat more delicate. According to part (v) of Theorem 1, every $x \in k^*$ may be written uniquely $x = \wp^n \varepsilon^k (1 + a_1\wp + a_2\wp^2 + \cdots)$. That is we have a direct product

$$k^* \cong \mathbb{Z} \times \mathbb{Z}_{q-1} \times A,$$

where $A = \{x\colon |x-1| < 1\}$ is a compact group. Therefore by duality theory

$$\hat{k}^* \cong T \times \mathbb{Z}_{q-1} \times D,$$

where D is the discrete group $D = \hat{A}$. Indeed for $\pi \in \hat{k}^*$, we can write $\pi = (\rho,\alpha,\theta)$, $\rho \in R/\mathbb{Z}$, $\alpha \in \mathbb{Z}_{q-1}$, $\theta \in D$ where if $x = \mathfrak{p}^n \varepsilon^k a$, then $\pi(x) = e^{2\pi i n\rho} e^{2\pi i \alpha k/q-1} \theta(a)$.

REMARK. Let $m > 0$ be an integer and $(k^*)^m$ = the group of mth power elements in k^*. Then $k^*/(k^*)^m \cong (\mathbb{Z}/\mathbb{Z}^m)(\mathbb{Z}_{q-1}/\mathbb{Z}^m_{q-1})(A/A^m)$, and this is a finite extension if and only if $[A: A^m] < \infty$. If $ch(k) = 0$, then that is the case (see e.g. Lang [1]). We shall have use for this observation later.

Now we consider algebraic groups in *characteristic zero*. Some of what we have to say holds more generally for algebraic varieties, but we shall say little about that. The basic references for this material are Borel [1,2] and Borel and Tits [1]. Let K be an algebraically closed field of characteristic zero. Then $GL(n,K)$ is an *affine* (i.e. Zariski-closed) subvariety of K^{n^2+1} by $g = (g_{ij}) \longleftrightarrow (g_{11}, g_{12}, \cdots, g_{nn}, (\det g)^{-1})$. Also we give $GL(n,K)$ the relative Z-topology as a subset of $M(n,K) \cong K^{n^2}$. It forms then an open subgroup. By a *linear algebraic group* G we mean a Z-closed subgroup of $GL(n,K)$, for some $n \geqq 1$;

$G = \{g = (g_{ij}) \in GL(n,K): p_\alpha(g_{ij}) = 0, \quad p_\alpha(\alpha \in J)$ a set of polynomials .

The ring of *regular functions* $K[G]$ on G is defined as follows:

$$K[G] = K[X_{ij},Y]/I, \quad I = \text{the polynomials vanishing on } G \text{ (as a subset of } K^{n^2+1}).$$

This is an *affine algebra* (i.e. is finitely generated and has no nilpotent elements). If G is connected then $K[G]$ is an integral domain, and its field of fractions $K(G)$ is the field of rational functions on G.

EXAMPLES. (1) If $G = G_a = \{\begin{pmatrix} 1 & x \\ 0 & 1 \end{pmatrix}: x \in K\}$, then $K[G] = K[x]$ = the usual polynomial ring in one indeterminate over K. Of course $K(G)$ = the field of rational functions in x.

(2) If $G = \{\begin{pmatrix} a & 0 \\ 0 & a^{-1} \end{pmatrix} : a \in K^*\}$, then $K[G] = K[a,a^{-1}]$ = the ring of all finite Laurent series in the indeterminate a. K(G) is the field of rational functions in a.

Now let $k \subseteq K$ be a subfield. We say G is *defined over* k (or G is a k-group), and write G def/k if I has a set of generators in $k[X_{ij},Y]$. Since the characteristic is zero, this is the same as assuming $p_\alpha \in k[X_{ij}]$. We write then $k[G] = k[X_{ij},Y]/I_k$, I_k = the polynomials with coefficients in k that vanish on G. The quotient field k(G) of k[G] is a subfield of K(G) consisting of rational functions defined over k. Also we set $G_k = GL(n,k) \cap G$. G_k is Z-closed in GL(n,k), and so is a locally compact group if k is a locally compact field.

We write G^0 for the neutral component of G (in the Z-topology). If $k = \mathbb{C}$ the two notions of connectivity are the same; but if $k = \mathbb{R}$, then $(G^0)_{\mathbb{R}}$ may not be connected in the usual topology (e.g. $G = \{\begin{pmatrix} \alpha & 0 \\ 0 & \alpha^{-1} \end{pmatrix} : \alpha \in \mathbb{C}^*\}$). It is always true that $[G: G^0] < \infty$.

The set $\mathfrak{g}$ of K-derivations of K[G] which commute with right translations is the Lie algebra of G. $\mathfrak{g}$ is canonically isomorphic to the tangent space $T(G)_e$ (see Borel [2]), and so G and G^0 have the same Lie algebra. If G is def/k, then $\mathfrak{g} = \mathfrak{g}_k \otimes_k K$ where $\mathfrak{g}_k$ = the derivations leaving k[G] stable. In the case of characteristic zero, the usual correspondences between groups and algebras are valid. G operates on itself via inner automorphisms, Int g: $x \to gxg^{-1}$. The differential of this map gives rise to the adjoint representation of G on $\mathfrak{g}$.

By a *homomorphism* $\rho: G \to G'$ we mean a group homomorphism such that $\rho^0: K[G'] \to K[G]$, $\rho^0(f)(g) = f(\rho(g))$, is an algebra homomorphism. If G and G' are def/k, we say ρ is def/k if $\rho^0: k[G'] \to k[G]$. The corresponding morphism $\rho_*: \mathfrak{g} \to \mathfrak{g}'$ is easily defined. A *rational representation* is a morphism $\rho: G \to GL(n,K)$. Then if $g \to (g_{11},\cdots,(\det g)^{-1})$ each $\rho(g)$ is a polynomial function of $g_{11},\cdots,(\det g)^{-1}$. By a *character* we mean a rational representation into $GL(1,K) \cong K^*$. Let $X(G)$ denote the set of characters.

If G is an algebraic group, H a closed subgroup, then G/H can be given the structure of an algebraic variety, and of a (not necessarily linear) algebraic group if H is also normal. To say that $G = HN$ is a semidirect product means in addition to the usual group-theoretic properties that $H \times N \to N$ is a morphism of algebraic varieties. If the group and the preceding morphism are defined over k, we say that $G = HN$ is def/k.

REMARK. We note in passing that much of the algebro-geometric structure we have been describing is defined more generally for affine varieties (or even algbraic varieties). Although we shall on a few occasions use the language of algebraic geometry applied to other than groups, we do not take the time to expound on that material here. See for example Borel [2, Chapter 0]. The reader is encouraged to think of an affine variety as a Z-closed subset of affine space, and a complete (in particular projective) variety as a compact manifold.

Every $g \in GL(n,K)$ is written uniquely as $g = g_s g_u$, where g_s is semisimple (i.e. conjugate to a diagonal matrix), g_u is unipotent, and $g_s g_u = g_u g_s$. If $G \subseteq GL(n,K)$ is algebraic and $g \in G$, then $g_u, g_s \in G$ also. Set $G_s = \{g \in G\colon g = g_s\}$, $G_u = \{g \in G\colon g = g_u\}$.

We continue our description of basic algebraic group structures with some information on tori, unipotent groups, solvable groups, semisimple and reductive groups, and Borel and parabolic groups.

DEFINITION. An *algebraic torus* is a group which is isomorphic to K^{*n}.

THEOREM 2. *For a connected algebraic group* G, *the following are equivalent:*

(i) G *is a torus.*

(ii) $G = G_s$.

(iii) G *is conjugate to a diagonal group.*

Let T be a torus. Then $x \in T$ can be represented $x = (x_1, \cdots, x_n)$, $x_1 \in K^*$. Hence a character χ of T can be written

$$\chi(x) = x_n^{m_1} \cdots x_n^{m_n}, \quad m_2 \in \mathbf{Z}.$$

In particular $X(T) \cong \mathbf{Z}^n$.

THEOREM 3. *Let* T *be* def/k. *The following are equivalent:*

(i) *Every* $\chi \in X(G)$ *is* def/k.

(ii) T *has a diagonal realization over* k, *i.e. there is a basis of* K^n *consisting of elements of* k^n *such that* T *is represented by diagonal elements with respect to that basis.*

(iii) *For every* $\rho : T \to GL(m,K)$, ρ def/k, *the group* $\rho(T)$ *is diagonalizable over* k.

When T satisfies the conditions of Theorem 3, we call T *k-split*. If $X(T)_k = \{1\}$, i.e. if there are no non-trivial characters def/k, we call T *anisotropic* over k. If T is def/k, then T is k-split $\Leftrightarrow T_k \cong k^{*n}$. If in addition k is a local field, then T is anisotropic over $k \Leftrightarrow T_k$ is compact.

THEOREM 4. *Let* T *be a* k-*torus. Then there are two uniquely defined* k-*subtori* T_a, T_d *such that* T_d *is* k-*split,* T_a *is anisotropic over* k *and* $T = T_a T_d$ *is an almost direct product* ($T_a \cap T_d$ *is finite).*

Another fact is that if $S \subseteqq T$ is a subtorus, there exists a torus S_1 such that $T = SS_1$ is a direct product.

EXAMPLES. (1) $T = \{\begin{pmatrix} \alpha & 0 \\ 0 & \alpha^{-1} \end{pmatrix} : \alpha \in \mathbb{C}^*\}$, $T_{\mathbb{R}} = \{\begin{pmatrix} a & 0 \\ 0 & a^{-1} \end{pmatrix} : a \in \mathbb{R}^*\}$.

(2) $T' = gTg^{-1}$, $g = \begin{pmatrix} 1 & i \\ i & 1 \end{pmatrix}$, $T' = \{\begin{pmatrix} \gamma & \delta \\ -\delta & \gamma \end{pmatrix} : \gamma^2+\delta^2 = 1\}$,

$T'_{\mathbb{R}} = \{\begin{pmatrix} \cos\theta & \sin\theta \\ -\sin\theta & \cos\theta \end{pmatrix} : \theta \in \mathbb{R}\}$.

(3) $T'' = g_1 T g_1^{-1}$, $g_1 = \begin{pmatrix} 1 & i \\ 0 & 1 \end{pmatrix}$, $T'' = \begin{pmatrix} \alpha & -i(\alpha-\alpha^{-1}) \\ 0 & \alpha^{-1} \end{pmatrix}$ is not

def/$\mathbb{R}$, $T''_{\mathbb{R}} = \begin{pmatrix} \pm 1 & 0 \\ 0 & \pm 1 \end{pmatrix}$.

Next G is called *unipotent* if $G = G_u$.

EXAMPLE. $G = \{\begin{pmatrix} 1 & & * \\ & \ddots & \\ 0 & & 1 \end{pmatrix} :$ entries in $K\}$.

In fact any connected unipotent group is conjugate to a subgroup of the above group in some dimension. Moreover (in characteristic zero), unipotent groups are automatically connected. We have

$$G = G_0 \supseteqq G_1 \supseteqq \cdots \supseteqq G_n = \{e\},$$

where $G_i/G_{i+1} \cong G_a$; and $\exp\colon \mathfrak{g} \to G$ is an isomorphism of algebraic varieties.

Next, an algebraic group is called *solvable* (resp. *nilpotent*) if it is so as an abstract group.

THEOREM 5. *Let G be connected and solvable.*

(i) *G is conjugate to a group of triangular matrices.*

(ii) *If G operates on a complete variety, it has a fixed point.*

(iii) *G_u is a connected normal subgroup and there is a maximal torus T in G such that $G = T \cdot U$. This semidirect product is* def/k *if G is. T is uniquely determined up to conjugacy by $(G_u)_k$.*

(iv) *There is a composition series $G = G_0 \supseteqq G_1 \supseteqq \cdots \supseteqq G_n = \{e\}$ where $G_i/G_{i+1} \cong G_a$ or $GL(1,K)$.*

(v) *G is nilpotent $\Longleftrightarrow$ T and G_u commute.*

Continuing, let G be an arbitrary algebraic group. By the *radical* $R(G)$ of G we mean the maximal connected normal solvable subgroup of G. Such a group in fact exists. Similarly one has the *unipotent radical* $R_u(G)$ = the maximal connected normal unipotent subgroup of G. G is called *semisimple* or *reductive* according as $R(G) = \{e\}$ or $R_u(G) = \{e\}$. Note $R(G) = R(G^0)$, $R_u(G) = R_u(G^0)$, and both are def/k.

The quotient $G/R(G)$ is semisimple, but $R(G)$ doesn't (quite) split in G. It is better to look at $R_u(G)$ which does split.

LEVI DECOMPOSITION (Mostow [1]). There exists a maximal reductive k-subgroup H of G (called a *Levi factor*) such that $G = H \cdot R_u(G)$ is a semidirect product. H is uniquely defined up to conjugation by $R_u(G)_k$.

A word of caution: $R_u(G) \subsetneq G_u$ in general.

THEOREM 6. *If* G *is an algebraic group, the following are equivalent:*

(i) G^0 *is reductive.*

(ii) $G^0 = S \cdot G'$ *is an almost direct product, where* $S = R(G^0)$ *is a central torus and* G' *is semisimple.*

(iii) G^0 *has a locally faithful fully reducible representation.*

(iv) *All rational representations of* G *are fully reducible.*

Note if G is connected and $G = H \cdot R_u(G)$ is a Levi decomposition, then $H = H'T$ where H' is connected semisimple and T is a torus. Then $R(G) = T \cdot R_u(G)$, and G is an almost semidirect product of H' and $R(G)$. Also any connected semisimple group is an almost direct product of connected simple groups, the latter meaning the only closed normal subgroups are finite groups.

THEOREM 7. *Let* G *be a connected algebraic group.*

(i) *All maximal tori are conjugate. Every semisimple element is contained in a torus.*

(ii) *All maximal connected solvable subgroups (called* Borel *groups) are conjugate. Every element of* G *belongs to one such group.*

(iii) *If* P *is a closed subgroup of* G, *then* G/P *is a projective variety* $\iff$ P *contains a Borel group. (Such* P's *are called* parabolic groups.*)*

By the *rank* of G we mean the dimension of a maximal torus. There always exist maximal tori def/k, but there may not exist Borel groups def/k.

EXAMPLES. (1) A typical unipotent group is $\begin{pmatrix} 1 & & * \\ & \ddots & \\ 0 & & 1 \end{pmatrix}$. There is no torus here, so the rank is zero.

(2) A typical solvable group is $G = \begin{pmatrix} * & & * \\ & \ddots & \\ 0 & & * \end{pmatrix}$. Here $R_u(G) = \begin{pmatrix} 1 & & * \\ & \ddots & \\ 0 & & 1 \end{pmatrix}$ and a maximal torus T of G would be the subgroup of diagonal matrices. $G = T \cdot R_u(G)$ and rank $G \equiv n$ if the matrices are $n \times n$.

(3) Semisimple group. Take $G = SL(n,K) = \{g \in GL(n,K) : \det g = 1\}$. A maximal torus is $T = \{\begin{pmatrix} a_1 & & 0 \\ & \ddots & \\ 0 & & a_n \end{pmatrix} : a_1 \cdots a_n = 1\}$, and a Borel group is $B = \begin{pmatrix} * & & * \\ & \ddots & \\ 0 & & * \end{pmatrix}$. The parabolic groups are of the form $P = \begin{pmatrix} \boxed{*} & & * \\ & \boxed{*} & \\ 0 & & \boxed{*} \end{pmatrix}$. The rank of G is n-1.

(4) Reductive group. Let $G = GL(n,K)$. Then the derived group G' is $SL(n,K)$, the central torus is $T = \{\begin{pmatrix} a & & 0 \\ & \ddots & \\ 0 & & a \end{pmatrix} : a \in K^*\}$. Note that $G' \cap T = \{\begin{pmatrix} a & & 0 \\ & \ddots & \\ 0 & & a \end{pmatrix} : a^n = 1\}$, a finite group. The rank of G is n.

EXERCISE. Find an example of a group G def/$\mathbb{R}$ such that $G_{\mathbb{R}} = SO(n)$.

Before going on to representations of semisimple p-adic groups, we close this section with a discussion of root systems. Let V be a finite-dimensional vector space over K with a positive-definite scalar product. A subset Φ of V is a *root system* when

(1) Φ is a finite set of non-zero vectors such that $\Phi = -\Phi$.

(2) for every $\alpha \in \Phi$, $s_\alpha(\Phi) = \Phi$ where s_α denotes the reflection through the hyperplane perpendicular to α.

(3) if $\alpha, \beta \in \Phi$, then $2(\alpha,\beta)/(\alpha,\alpha) \in \mathbb{Z}$.

The group generated by the s_α: $\alpha \in \Phi$ is a finite group called the *Weyl group*. The integers $2(\alpha,\beta)/(\alpha,\alpha)$ are called the *Cartan integers* of Φ. It is a simple algebraic fact that

$$s_\alpha(v) = v - 2\alpha(\alpha,v)/(\alpha,\alpha).$$

Hence $s_\alpha(\beta)-\beta$ is an integral multiple of α.

A root system Φ in V is the direct sum of $\Phi' \subseteq V'$ and $\Phi'' \subseteq V''$ if $V = V' \oplus V''$ and $\Phi = \Phi' \cup \Phi''$. Φ is called *irreducible* if it is not the direct sum of two subsystems. Then every root system is the direct sum of irreducible subsystems. Moreover if $\alpha \in \Phi$ and $\lambda\alpha \in \Phi$, then $\lambda = \pm 1, \pm 2, \pm\frac{1}{2}$. Φ is called *reduced* if $\forall \alpha \in \Phi$, $\lambda\alpha \in \Phi \Rightarrow \lambda = \pm 1$.

THEOREM 8. *The only reduced irreducible root systems are:*

$$A_n (n \geqq 1),\ B_n (n \geqq 2),\ C_n (n \geqq 3),\ D_n (n \geqq 4),\ E_6,\ E_7,\ E_8,\ F_4,\ G_2.$$

The reader is encouraged to consult Helgason [1] and Humphreys [1] for an elaboration on these root systems and the groups that correspond to them.

A hyperplane of V is called *singular* if it is orthogonal to a root. A *Weyl chamber* is a connected component of the complement of the union of all the singular hyperplanes. Fixing a chamber C determines an ordering on the roots by $\alpha > 0$ if $(\alpha,v) > 0$ all $v \in C$. A root is *simple* if it is positive and not the sum of two other positive roots. We write Δ for the simple roots. Δ is called *connected* if it cannot be written as a disjoint union of orthogonal subsets.

THEOREM 9. (i) W *acts simply transitively on the Weyl chambers.*

(ii) *Every* $\alpha \in \Phi$ *is an integral linear combination of simple roots, all coefficients having the same sign.*

(iii) Φ *is irreducible* $\iff$ Δ *is connected.*

Now let G be a semisimple (or more generally reductive) algebraic group. Let $T \subseteq G$ be a maximal torus. G operates on $\mathfrak{g}$ by the adjoint representation and of course $\mathrm{Ad}_G T$ is completely diagonalizable. Therefore

$$\mathfrak{g} = \mathfrak{g}_1^T \oplus \sum_{\alpha \neq 1} \mathfrak{g}_\alpha^T$$

where $\mathfrak{g}_\alpha^T = \{X \in \mathfrak{g} : \mathrm{Ad}\ t(X) = \alpha(t)X\}$, $\alpha \in X(T)$. The set $\Phi = \{\alpha : \mathfrak{g}_\alpha^T \neq \{0\},\ \alpha \neq 1\}$ is a root system. If T is maximal, the set Φ of roots obtained are called the roots of G. This coincides in case $K = \mathbb{C}$ with the root systems investigated in Chapter I.

Finally what happens in the relative case $k \subseteq K$. Let G be def/k. Then the maximal k-split tori are conjugate over k. Let S be one of them; $\dim S$ = k-rank of G. G is called *anisotropic* if and only if its k-rank is zero. G has a parabolic subgroup def/k $\iff \mathrm{rank}_k\, G > 0$. If k is a local field, then G is anisotropic $\iff G_k$ is compact. Now $Z(S)$ is a reductive group, and in fact is the neutral component of $N(S)$. The finite group $W_k(G) = N(S)/Z(S)$ is called the Weyl group of G relative to k.

The elements of $\Phi(G,S)$ are called the k-roots, or k-restricted roots. Write $\Phi_k(G)$ for $\Phi(G,S)$. Of course $\Phi_K(G) = \Phi(G)$ as before. If G is simple, then $\Phi_k(G)$ is irreducible, but in general Φ_k is not reduced unless k is algebraically closed.

All the minimal parabolic k-groups are conjugate. Furthermore there is a maximal k-split torus S and a minimal k-parabolic P such that $P = Z(S)R_u(P)$. Put $U = R_u(P)$. Then we have

THEOREM 10. (Bruhat Decomposition - Borel [1]).
$G_k = U_k N(S)_k U_k = \bigcup_{w \in W_k} U_k\, w\, P_k$, *a disjoint union.*

Finally we give some information on the standard parabolic groups. Let G be def/k, P a minimal k-parabolic, S a maximal k-split torus in P, $\Phi_k(G)$ the k-roots. There is a well-determined positive Weyl chamber (and thus a choice of positive roots) such that $U = R_u(P) = \exp \sum_{\alpha>0} \mathfrak{g}^S_\alpha$. Let Δ_k be a set of simple roots for this ordering. Fix a subset Θ of Δ_k. Set S_Θ = the identity component of $\bigcap_{\alpha\in\Theta} \ker \alpha$. S_Θ is a k-split torus whose dimension is $\mathrm{rank}_k(G)$ - card Θ. Then define the *standard parabolic* group P_Θ to be the subgroup generated by $Z(S_\Theta)$ and U. In fact P_Θ has a Levi decomposition $P_\Theta = Z(S_\Theta)U_\Theta$, where $U_\Theta = R_u(P_\Theta) = \exp \sum \mathfrak{g}^S_\alpha$, the sum going over all positive roots that are not linear combinations of elements in Θ.

THEOREM 11. (Borel [1]). (i) *Every parabolic* k-*subgroup is conjugate over* k *to one and only one of the standard parabolic* k-*subgroups.*

(ii) *Let* W_Θ *be the subgroup of* W_k *generated by the reflections* s_α: $\alpha \in \Theta$. *Then if* Θ *and* Θ' *are two subsets of* Δ_k, *we have*

$$(P_\Theta)_k \backslash G_k / (P_{\Theta'})_k \cong W_\Theta \backslash W_k / W_{\Theta'}.$$

EXERCISE. Compute the Bruhat decomposition and the P_Θ's for $G = SL(n,K)$.

In the remainder of this chapter we shall commit on occasion an abuse of terminology by referring to groups of rational points G_k as algebraic groups.

B. REPRESENTATIONS OF SEMISIMPLE p-ADIC GROUPS

Almost everything we have to say in this section can be found in Harish-Chandra [10, 12]. We continue writing k for a locally compact field of characteristic zero, K an algebraically closed field $\supseteq k$. Let G be a connected semisimple algebraic group def/k. Suppose P is a k-parabolic subgroup. Let N denote the unipotent radical of P. Then N is def/k and P is the normalizer of N in G. We can choose a Levi subgroup M of P such that $P = MN$ and $(m,n) \to mn$, $M \times N \to P$, is a k-isomorphism of algebraic varieties. The groups P and M are connected. Hence M is an almost direct product of its semisimple derived group $\mathfrak{D}_M$ and its toral center Z_M. Let A be the maximal k-split torus in Z_M. Then M is the centralizer of A in G. We call A a *split component* of P. Then for any split component A' of P there exists u N_k such that $A' = uAu^{-1}$. Hence dim A depends only on P and is called the *parabolic rank* of P.

Next let $P_k = P \cap G_k$. Then P_k is called a parabolic subgroup of G_k. The group $A_k = A \cap G_k$ is called a split component of P_k. Finally the locally compact group P_k is a semidirect product $P_k = M_k N_k$, $M_k = M \cap G_k$, $N_k = N \cap G_k$ (see section D). The pair (P_k, A_k) is often called a *parabolic pair*.

It is possible to talk about restricted roots as in the real case. By a (restricted) root α of P_k (or (P_k, A_k)) we mean a non-zero $\alpha \in \mathfrak{a}_k^* = \mathrm{Hom}_k(\mathfrak{a}_k, k)$ with the property that $\mathfrak{g}_k^\alpha = \{X \in \mathfrak{g}_k \colon [X,H] = \alpha(H)X,\ H \in \mathfrak{a}_k\} \neq \{0\}$. Then of course

$$\mathfrak{g}_k = \sum_{\alpha>0} \mathfrak{g}_k^\alpha + \sum_{\alpha<0} \mathfrak{g}_k^\alpha + \mathfrak{g}_k^0 \quad \text{where} \quad \mathfrak{g}_k^0 = \mathfrak{m}_k + \mathfrak{a}_k.$$

So far things look very much like the Lie group case. But it would be misleading if we left the impression that the theory is a simple matter of translation of results. That is far from the case.

For one thing (when k is non-archmidean), since G_k is a totally disconnected group, it has many compact-open subgroups. In fact, it has maximal compact-open subgroups -- but unlike the real case, these do not have to be all conjugate. It is possible to describe the conjugacy classes of such maximal compacts in terms of a so-called Iwahori subgroup, but we prefer not to go into that here. However there is a certain good *choice* of a maximal compact-open subgroup. In order to explain that, we first need some more terminology. In the following k is understood to be non-archmidean. It is convenient (and hopefully not too confusing) to drop the k henceforth in the subscripts: G denotes the group of rational points.

Let (P_i, A_i), $i = 1,2$, be two parabolic pairs. Write $(P_1, A_1) > (P_2, A_2)$ if $P_1 \supseteq P_2$ and $A_1 \subseteq A_2$. A pair is called *minimal* if it is minimal with respect to this order. Let $w(A_1, A_2) = \{s: A_1 \to A_2,$ bijection, $s(a) = yay^{-1}$ some $y \in G\}$. This is a finite set. For $A_1 = A_2 = A$, write $w(A)$. Now let (P_0, A_0) be minimal, with $P_0 = M_0 N_0$ the corresponding Levi decomposition. For α a root of (P_0, A_0) denote by ξ_α the character of A_0 obtained by lifting α. Set $A_0^+ = \{a \in A_0 : |\xi_\alpha(a)| \geqq 1 \ \forall \alpha\}$.

THEOREM 1. (Bruhat-Tits). *There exists a choice of maximal compact-open subgroup* K *of* G *such that:*

(i) $G = KP_0$.

(ii) *There is a finite subset* ω_{M_0} *of* M_0 *such that* $G = KA_0^+ \omega_{M_0} K$.

(iii) *Every element of* $w(A_0)$ *has a representative in* K.

(iv) *If* $(P, A) > (P_0, A_0)$ *and* $P = MN$ *then* $P \cap K = (M \cap K)(N \cap K)$.

(v) *Put* $K_M = K \cap M$, ${}^*P_0 = M \cap P_0$. *If we replace* (G, P, A_0, K) *by* $(M, {}^*P_0, A_0, K_M)$, *then* (i)-(iv) *are satisfied for the latter.*

Note (v) allows for the induction that occurs in many of Harish-Chandra's arguments.

EXAMPLE. $k = \mathbb{Q}_p$, $G = SL(n,k)$. Then for K we may select the group $SL(n,\mathcal{O})$, $\mathcal{O}$ = the ring of p-adic integers in k. Also $P_0 = \begin{pmatrix} * & & * \\ & \ddots & \\ 0 & & * \end{pmatrix}$.

Now to talk about the representation theory of G, it is necessary to do things somewhat differently from the real case. First of all we observe that everything said so far holds as well for reductive groups. Then we have to discuss the notion of admissibility. For this it is convenient to speak momentarily about general locally compact totally-disconnected groups. For such groups G, we write $C_0^\infty(G)$ for the locally constant compactly supported functions. Call these *smooth functions*. If V is a complex vector space, $C_0^\infty(G;V)$ denotes the smooth V-valued functions. By a *smooth representation* of G in V we mean a representation $G \to GL(V)$ such that $x \to x\cdot v$, $G \to V$, is smooth for all $v \in V$. If $H \subseteq G$ is a closed subgroup and σ is a smooth representation of H on V, we can define a smooth representation $\pi = \mathrm{Ind}_H^G \sigma$ as follows. Let $V(\pi)$ be the smooth functions $\beta\colon G \to V$ satisfying (1) $\beta(hx) = \sigma(h)\beta(x)$, $h \in H$, $x \in G$ and (2) β is compactly supported modulo H. Then π acts in $V(\pi)$ via $\pi(x)\beta(y) = \beta(yx)$, $\beta \in V(\pi)$.

A representation of G on V is called *admissible* if (1) π is smooth and (2) for any open subgroup H of G, $\dim V_H < \infty$, $V_H = \{v \in V\colon \pi(h)v = v \text{ all } h \in H\}$. An admissible representation π can be lifted to $C_0^\infty(G)$ as follows. Let $f \in C_0^\infty(G)$, $v \in V$. Choose a compact-open subgroup U such that $f(gu) = f(g)$, $u \in U$, $g \in G$, and $\pi(u)v = v$, $u \in U$. Then $\mathrm{Supp}\, f = \bigcup_{i=1}^{n} k_i U$. Hence

$$\pi(f)v = \int_G f(g)\pi(g)v\, dg = \int_{G/U}\int_U f(gu)\pi(gu)v\, du\, d\bar{g}$$

$$= \sum_{i=1}^{n} f(k_i)\pi(h_i)v \int_U du.$$

The vector $\pi(f)v$ thus obtained is independent of the choice of U.

EXERCISE. Show that π is admissible $\iff$
(1) $\forall v \in V$, there is $f \in C_0^\infty(G)$ such that $\pi(f)v = v$; and
(2) $\forall f \in C_0^\infty(G)$, $\pi(f)$ has finite range.

If V^* is the linear space dual of V, the dual representation π^* is defined by $\langle\pi^*(x)\lambda, v\rangle = \langle\lambda, \pi(x^{-1})v\rangle$, $\lambda \in V^*$, $v \in V$. Set $\tilde{V} = \{\lambda \in V^*: x \to \pi^*(x)\lambda$ is smooth$\}$, and $\tilde{\pi} = \pi^*|_{\tilde{V}}$. This is a smooth representation called the *contragradient* of π. We have π admissible $\iff$ $\tilde{\pi}$ is admissible, and $\tilde{\tilde{\pi}} = \pi$.

LEMMA 2. *Let* $H \subseteq G$ *be a closed subgroup,* σ *an admissible representation of* H. *Then* $\pi = \mathrm{Ind}_H^G \sigma$ *is an admissible representation of* G.

Two representations π_1, π_2 are called *equivalent* if there is a linear bijection T between their spaces that intertwines, $T\pi_1(x) = \pi_2(x)T$, $x \in G$. We denote $\mathcal{E}_c(G)$ = the set of equivalence classes of irreducible admissible representations of G. We also denote by $\mathcal{E}(G)$ the subset consisting of unitary classes, $\mathcal{E}(G) = \mathcal{E}_c(G) \cap \hat{G}$.

CONJECTURE 3. *If* G *is a reductive* k*-group, then every irreducible unitary representation is admissible, that is* $\hat{G} = \mathcal{E}(G)$.

COROLLARY. If Conjecture 3 holds, then reductive k-groups are CCR. Of course this is true if $k = \mathbb{R}$ or $\mathbb{C}$. For k non-archmidean, this is known only for SL(2) or GL(2) (See Jacquet and Langlands [1] -- there has also been considerable work on GL(n), see Howe [1]).

Next we need the notion of a supercuspidal representation.

DEFINITION. A continuous function f in G is said to be a *supercusp* form if

(1) Supp f is compact mod $Z = Z_G$,

(2) $f^P = 0$ for all parabolic subgroups $P \neq G$, where

$$f^P(x) = \int_N f(xu)\,du, \qquad P = MN.$$

Note the integral makes sense since $Z \subseteq M$.

A representation π is called *supercuspidal* if for every $\phi \in V$, $\tilde{\phi} \in \tilde{V}$, the function $x \to \langle \pi(x)\phi, \tilde{\phi}\rangle$ is a supercusp form. Let $°\mathcal{E}_c(G)$ denote all supercuspidal classes in $\mathcal{E}_c(G)$, and set $°\mathcal{E}(G) = °\mathcal{E}_c(G) \cap \hat{G}$. We also denote by $\hat{G}_{\mathcal{D}}$ the relative discrete series of G, that is the irreducible unitary representations which are square-integrable mod Z:

$$\hat{G}_{\mathcal{D}} = \{\pi \in \hat{G}\colon \text{ For some } \xi, \eta \in \mathcal{H}_\pi,$$

$$x \to (\pi(x)\xi, \eta) \text{ is square-integrable mod } Z\}.$$

Note that $°\mathcal{E}(G) \subseteq \hat{G}_{\mathcal{D}}$, and that $\hat{G}_d = \hat{G}_{\mathcal{D}}$ if Z is compact. The following result generalizes Theorems 3 and 4 of Chapter I, section B2.

THEOREM 4. *Let* G *be unimodular. Suppose* $\pi, \pi' \in \hat{G}$, $\pi|_Z = \chi I$, $\pi'|_Z = \chi' I$, $\chi, \chi' \in \hat{Z}$.

(a) *The following are equivalent:*

(1) $\pi \in \hat{G}_{\mathcal{D}}$.

(2) $\int_{G/Z} |(\pi(g)\phi, \psi)|^2 \, d\bar{g} < \infty \ \forall \phi, \psi \in \mathcal{H}_\pi$.

(3) π *is equivalent to a subrepresentation of* $\mathrm{Ind}_Z^G \chi$.

If (1)-(3) *hold, there is a number* $d(\pi)$ *such that*

$$\int_{G/Z} \overline{(\phi_1, \pi(x)\psi_1)}\, (\phi_2, \pi(x)\psi_2)\, dx = d(\pi)^{-1}\, \overline{(\phi_1, \phi_2)}(\psi_1, \psi_2)$$

(b) *If* π *and* π' *are in* $\hat{G}_{\mathcal{D}}$, *but are not equivalent, then*

$$\int_{G/Z} \overline{(\phi,\pi(x)\psi)}(\phi',\pi'(x)\psi')dx = 0.$$

REMARK. The author believes he has heard that Conjecture 3 would follow if one knew that $\inf_{\pi \in {}^\circ\mathcal{E}(G)} d(\pi) > 0$.

Now we are ready for the definition and main properties of the principal series. G is of course again (the k-rational points of) a reductive k-group. Let (P_i,A_i), $i = 1,2$ be parabolic pairs, σ_i admissible representations of M_i, Δ_{P_i} the modular functions of P_i. Set

$$\pi_i = \mathrm{Ind}_{P_i}^{G} (\Delta_{P_i}^{\frac{1}{2}} \sigma_i).$$

THEOREM 5. (i) *Assume* σ_1 *and* σ_2 *are both supercuspidal. Then* $\mathcal{I}(\pi_1,\pi_2) = 0$ *unless* A_1 *and* A_2 *are conjugate* (P_1 *and* P_2 *associate*).

(ii) *Suppose* $P_1 = P_2 = P$, *etc. Suppose* σ_1, σ_2 *are irreducible and supercuspidal representations of* M, *i.e.* $\sigma_i \in {}^\circ\mathcal{E}_c(M)$. *Then*

$$\mathcal{I}(\pi_1,\pi_2) \leqq \# s \in w(A): s\sigma_1 \cong \sigma_2$$

(iii) *For every* $\sigma \in {}^\circ\mathcal{E}(M)$ *such that* $s\sigma \not\cong \sigma$, $s \in w(A)$, $s \neq 1$, *the representation* $\pi = \mathrm{Ind}_P^G \Delta_P^{\frac{1}{2}} \sigma$ *is unitary, admissible, and irreducible.*

The latter are called the *principal series* corresponding to P. In a sense the representation theory of G can be reduced to the study of supercuspidals. We shall try to make that precise, but first we need some more terminology.

If π is admissible in V, we use $\mathfrak{A}(\pi)$ to denote the $\mathbb{C}$-span of all matrix coefficients $x \to \langle\pi(x)v,\tilde{v}\rangle$, $v \in V$, $\tilde{v} \in \tilde{V}$. We set $\mathfrak{A}(G) = \bigcup_\pi \mathfrak{A}(\pi)$ as π runs over all admissible representations. Then for any parabolic pair (P,A) and $f \in \mathfrak{A}(G)$, there is exactly one element $f_P \in \mathfrak{A}(M)$ with the following property: Given a compact set S in M there exists $t \geq 1$ such that

$$\Delta^{\frac{1}{2}}_P(ma)f(ma) = f_P(ma),$$

for $m \in S$, $a \in A^+(t) = \{a \in A: |\xi_\alpha(a)|_P \geq t$ for every simple root α of $(P,A)\}$.

Furthermore, there is a direct sum

$$\mathfrak{A}(G) = \sum_{\chi \in X(Z)} \mathfrak{A}(G,\chi)$$

where $X(Z) = \{\chi: Z \to C^*,\ \chi$ a continuous homorphism$\}$

$$\mathfrak{A}(G,\chi) = \{f \in \mathfrak{A}(G): (\rho(z)-\chi(z))^n f = 0 \text{ for some } n\},$$

ρ = the right regular representation of G. If $f \in \mathfrak{A}(G)$, we write $f_{P,\chi}$ for the component of f_P in $\mathfrak{A}(M,\chi)$. Finally, if π is admissible, we set $\mathcal{X}_\pi(P,A) = \{\chi: f_{P,\chi} \neq 0\}$. It is known that if $\pi \in \mathcal{E}_c(G)$, then $\mathcal{X}_\pi(P,A)$ is a finite set which is empty if $\pi \in {}^\circ\mathcal{E}_c(G)$ and $P \neq G$.

Let $\pi \in \mathcal{E}_c(G)$. We can always choose P to be π-minimal, that is $\mathcal{X}_\pi(P,A) \neq \emptyset$ but $\mathcal{X}_\pi(P',A') = \emptyset$ if $(P',A') < (P,A)$. Let $\overline{P}$ be the opposed parabolic (obtained by reversing the roots). Then we have

THEOREM 6. *Let* $\pi \in \mathcal{E}_c(G)$, P *π-minimal,* $\chi \in \mathcal{X}_\pi(P,A)$. *Then there exists an irreducible, admissible supercuspidal representation* σ *of* M *such that*

(i) $\sigma|_Z = \chi$

(ii) π *is a subrepresentation of* $\mathrm{Ind}_{\overline{P}}^G\, \Delta_{\overline{P}}^{\frac{1}{2}}\sigma$.

If χ *is unitary, then so is* π.

CAUTION. There are representations of $\hat{M}_{\mathcal{D}}$ which are not in ${}^\circ\mathcal{E}(M)$ -- these are the so-called *special representations*. We do not devote any space to that here.

Thus to a tremendous extent we see that the study of representations of G is reduced to the study of supercuspidal representations.

Unfortunately there is not yet a complete theory of these analogous to the discrete series for real groups. This is in part due to the existence of non-conjugate maximal compacts and more particularly to the existence of non-conjugate compact Cartan subgroups. As might be expected, one cannot compute these representations by any well-determined prescription (except in certain low-dimensional cases); and unfortunately because of the previously cited difficulties, one does not have a theory of their characters as in the real case. However there are some results.

Indeed, note that if π is admissible, then $\pi(f)$ is of finite rank for $f \in C_0^\infty(G)$. Hence $\Theta_\pi: f \to \operatorname{Tr} \pi(f)$ is a distribution.

THEOREM 7. (i) (Harish-Chandra [10]) *If* $\pi \in {}^\circ\mathcal{E}(G)$, *then* Θ_π *is a locally integrable function on* G, *which is locally constant in the set of regular elements.*

(ii) (Van Dyk [1]) *If* $\pi = \operatorname{Ind}_P^G \Delta_P^{\frac{1}{2}}\sigma$, $\sigma \in {}^\circ\mathcal{E}(M)$, *then it is possible to compute* Θ_π *(by realizing* $\pi(f)$ *as a kernel operator on* $L_2(K,\mathcal{H}_\sigma)$ *and integrating down the diagonal) in terms of* Θ_σ. *The result is in analogy with Chapter I, Theorem C4.*

Now let $\mathcal{E}'(G) = \{\pi \in \mathcal{E}(G): \mathcal{X}_\pi(P,A) \cap \hat{A} = \emptyset,\ P \neq G\}$. Then $\hat{G}_{\mathcal{D}} \subseteqq \mathcal{E}'(G)$. By considerations involving the Schwartz space, it is possible to generalize the results of Theorem 5. In particular one can prove

THEOREM 8. *Fix* $\sigma \in \mathcal{E}'(M)$ *and assume* $s\sigma \not\approx \sigma$, $s \in w(A)$, $s \neq 1$. *Then* $\pi = \operatorname{Ind}_P^G \Delta_P^{\frac{1}{2}} \sigma$ *is unitary, irreducible and admissible.*

Finally there is a provisional (preliminary) form of the Plancherel formula on G (which reduces matters to knowing the formal degrees on M). See Harish-Chandra [12] for that. As for examples, we are content to refer the reader to Jacquet and Langlands [1] and Sally and Shalika [1, 2].

C. REPRESENTATIONS OF UNIPOTENT p-ADIC GROUPS

In this section we give an account of Moore's results (Moore [2]) on the representation theory of unipotent p-adic groups. Let G be a unipotent (and so connected) algebraic group def/k, k a non-archimedean local field of characteristic zero. It is no loss of generality to assume $G \subseteq U = \{\begin{pmatrix} 1 & & * \\ & \ddots & \\ 0 & & 1 \end{pmatrix}$: entries in $K\}$. Let $\mathfrak{u} = \begin{pmatrix} 0 & & * \\ & \ddots & \\ 0 & & 0 \end{pmatrix} = LA(U)$. Then the exponential map $\exp\colon \mathfrak{u} \to U$ is a complete isomorphism of algebraic varieties. The inverse map is denoted $\log$ and we write $\mathfrak{g} = \log(G) = LA(G)$. Clearly $\mathfrak{g}$ is def/k and $\exp \mathfrak{g}_k = G_k$.

Now we change notation once again. Write N for G_k, $\mathfrak{n}$ for $\mathfrak{g}_k$, $\exp\colon \mathfrak{n} \to N$ for the exponential map. The claim is that the Kirillov description of the space $\hat{N}$ carries over without change to this setting. In order to proceed with the details we first need an analog of *exp* $2\pi i$.

Let $\varepsilon \in \hat{k}$ be a non-trivial character (Moore picks a special one, but that will not be essential for the presentation here). Then for every $x \in k$, $\varepsilon_x\colon y \to \varepsilon(xy)$ is a character and $x \to \varepsilon_x$ is a topological isomorphism of k onto $\hat{k}$. This can be reworded as follows. Let V be a finite-dimensional vector space over k, and let $f \in V^* = \mathrm{Hom}_k(V,k)$. Then the map

$$\Phi(f)\colon x \to \varepsilon(f(x))$$

is a character of the locally compact abelian group V. It is then a simple exercise to verify that $f \to \Phi(f)$ is an isomorphism of V^* onto $\hat{V}$.

Next let $V = \mathfrak{n}$, $f \in \mathfrak{n}^*$. By a *subordinate subalgebra* for f we mean a Lie subalgebra $\mathfrak{m}$ of $\mathfrak{n}$ such that $f([\mathfrak{m},\mathfrak{m}]) = 0$.

Let $\mathfrak{m}_f$ denote a maximal subordinate subalgebra for f. Then we can set $M_f = \exp \mathfrak{m}_f$ the corresponding group. Clearly the mapping

$$n \to \varepsilon(f(\log n)) = \Phi(f)(\log n), \quad N \to \mathbb{T}$$

is a continuous map. Moreover it is clear that the restriction of this map to M_f is a character χ_f of M_f (since $f([\mathfrak{m}_f,\mathfrak{m}_f]) = 0$). Set $\rho(f,M_f) = \mathrm{Ind}_{M_f}^G \chi_f$.

THEOREM 1. (Moore [2]) *The representation* $\rho(f,M_f)$ *is irreducible and independent of the choice of* M_f*, denote it* $\rho(f)$*. If* $g \in \mathfrak{n}^*$ *also, then* $\rho(f) \cong \rho(g)$ *if and only if there is* $n \in N$ *such that* $\mathrm{Ad}^*(n)f = g$*. Finally every element of* $\hat{N}$ *is of the form* $\rho(f)$ *for some* $f \in \mathfrak{n}^*$*. Thus the map* $\mathfrak{n}^*/N \to \hat{N}$ *is a bijection.*

THEOREM 2. (Moore [2]) N *is* CCR.

Both of these are proven by arguments analogous to the real case -- Theorem 1 by duplicating Kirillov's arguments, Theorem 2 by duplicating Fell's [2].

The result on $\dim \mathfrak{m}_f$ (Chapter IV Theorem A7) goes thorugh since it involves only elementary linear algebra. The results on the enveloping algebra (Chapter IV Theorem A8) do not generalize. The author unfortunately does not know whether results on characters and the Plancherel measure analogous to the real case have been investigated.

EXAMPLE. $$N = \left\{ \begin{pmatrix} 1 & x_1 \cdots x_n & z \\ & 1 \ddots 0 & y_1 \\ & \quad \ddots & \vdots \\ & 0 \quad 1 & y_n \\ & & 1 \end{pmatrix} : x_i, y_i, z \in k \right\}.$$

This is a two-step unipotent group. Essentially the same description of the representations as in Chapter IV, Example 1, p. 101 obtains. There are two families of representations: the representations trivial on $Z = Z_N$, i.e. characters of $N/Z = k^2$; and the one-parameter family π_ζ, $\zeta \in k^*$, where π_ζ is the unique (infinite-dimensional) irreducible whose restriction to Z acts via the non-trivial character $z \to \varepsilon(\zeta z)$.

D. ALGEBRAIC GROUPS AS GROUP EXTENSIONS

The basic idea we pursue here is the following. G is an algebraic group def/k, G_k = the locally compact group of rational points, ch(k) = 0. We know from the Levi decomposition that $G = HU$, $U = R_u(G)$, H a Levi factor. Claim: If we set $H_k = H \cap G_k$, $U_k = U \cap G_k$, then we have a semidirect product of locally compact groups $G_k = H_k U_k$ (to be justified momentarily in Lemma 1). The goal is then to employ the Mackey extension procedure and known results from unipotent and reductive groups to see what can be said about the representation theory of G_k. Results on this topic are still quite primitive at this time.

In the following we write $G = HU$ and Z is a maximal k-split torus contained in H.

Let Γ be the Galois group Gal(K/k), i.e. the group of automorphisms of the field K leaving k pointwise fixed. Clearly Γ acts on the group $GL(n,K)$ as a group of automorphisms and the fixed point set of Γ is precisely $GL(n,k)$. Now if G is an algebraic group, the same result is true. That is, Γ acts by automorphisms on G and the fixed point set is precisely G_k.

LEMMA 1. (Lipsman [5]) (i) G_k *is a semidirect product* $G_k = H_k U_k$.

(ii) *If* $\dim Z = m$, *then* $Z_k \cong (k^*)^m$.

(iii) *If* U *is abelian and* $\dim U = r$, *then* $U_k \cong k^r$.

Proof. (1) Clearly $U_k = U \cap G_k$ is closed and normal in G_k, and $H_kU_k \subseteq G_k$. Conversely, let $g \in G_k$. Then there are unique elements $h \in H$, $u \in U$ such that $g = hu$. Then for any $\sigma \in \Gamma$ we have $hu = g = g^\sigma = h^\sigma u^\sigma$. Now the fact that H and U are def/k translates precisely into the statement that H, U are preserved setwise by Γ. Hence $h^\sigma \in H$, $u^\sigma \in U$. But then by the uniqueness of the decomposition $G = HU$, we have $h = h^\sigma$, $u = u^\sigma$, i.e. $h \in H_k$, $u \in U_k$. That proves (1).

Statements (ii) and (iii) follow from the observations $Z \cong (K^*)^m$, $U \cong K^r$ (the latter by choosing a basis $X_1,\cdots X_r$ of $\mathfrak{n}$ and mapping $(x_1,\cdots,x_r) \to \exp(\sum_{i=1}^{r} x_iX_i)$, $x_i \in K$), together with the following.

LEMMA 2. (Lipsman [5]) *Let* G^1, G^2 *be algebraic groups* def/k *and suppose* $\alpha: G^1 \to G^2$ *is a group morphism* def/k. *Then* $\alpha(G^1_k) \subseteq G^2_k$. *If in addition* α *is bijective, then* $\alpha(G^1_k) = G^2_k$.

Proof. The Galois group Γ also acts on $\mathrm{Hom}(G^1,G^2)$ via $(\alpha^\sigma)(g) = [\alpha(g^{\sigma^{-1}})]^\sigma$, $\sigma \in \Gamma$. To say that α is defined over k is to say precisely that $\alpha^\sigma = \alpha$. Then by the computation: $g \in G^1_k \Rightarrow [\alpha(g)]^\sigma = \alpha^\sigma(g^\sigma) = \alpha(g) \Rightarrow g \in G^2_k$, we obtain the first part of the lemma. If in addition α is bijective, we reason as follows. Let $g_2 \in G^2_k$. Then by surjectivity, there is $g_1 \in G^1$ such that $\alpha(g_1) = g_2$. But for $\sigma \in \Gamma$, we have $\alpha(g_1^\sigma) = g_2^\sigma = g_2 = \alpha(g_1)$. Since α is injective, $g_1 = g_1^\sigma$. Hence $g_1 \in G^1_k$.

LEMMA 3. (Lipsman [5]) *The group* U_k *is type I and regularly embedded in* G_k.

Proof. (Sketch) We already know that U_k is even CCR. That it is type I follows. It is regularly embedded by the following reasoning. Suppose we knew that $\mathfrak{n}^*_k/G_k$ was countably separated. Then we use the map

$$(\mathfrak{u}_k^*/G_k)/(G_k/U_k) \to \mathfrak{u}^*/G_k,$$

which is easily seen to be a Borel isomorphism to deduce that $(\mathfrak{u}_k^*/U_k)/(G_k/U_k)$ is countably separated. But since the Kirillov theory applies, $\mathfrak{u}_k^*/U_k \to \hat{U}_k$ is a Borel isomorphism. Hence $\hat{U}_k/(G_k/U_k) \cong \hat{U}_k/G_k$ is countably separated. Without going into the details, we just say that $\mathfrak{u}_k^*/G_k$ is countably separated because for linear algebraic actions the orbits are always smooth varieties (see e.g. Borel [2]), in particular locally closed in the ordinary topology. Hence by results of Effros [1], the space $\mathfrak{u}_k^*/G_k$ is countably separated.

Note we also have the following famous

CONJECTURE 4. G_k *is type I.*

In fact Dixmier [1] proved this for k archimedean. His proof in the general case reduces the problem to the semisimple case, and so Conjecture 4 would follow if we knew that irreducible unitary representations of semisimple p-adic groups were admissible (section B, Conjecture 3).

Now we want to apply the Mackey [6] and Kleppner and Lipsman [1, 2] extension procedure to $U_k \subseteqq G_k$ in order to compute the representation theory and Plancherel measure of G_k. For this we need to compile data on (1) the H_k orbit structure on $\hat{U}_k$ and (2) the stability groups $(G_k)_\gamma$, $\gamma \in \hat{U}_k$, that arise. One has a substantial amount of information on the former, but not nearly enough on the latter.

Let us begin by restricting ourselves to the case U is abelian. If we let $\mathfrak{u}$ be the Lie algebra of U, then the exponential map effects a complete isomorphism of $\mathfrak{u}$ and U. Letting $V = \mathrm{Hom}_K(\mathfrak{u},K)$, we see that problems (1) and (2) can be formulated within the more general context of studying finite-dimensional representations

$$H \to GL(V)$$

$$H_k \to GL(V_k).$$

In this vein, there is a beautiful result of R. Richardson which is applicable. To state it we need a definition. If G is a topological group and X a G-set (i.e. a topological space on which G acts), we say there exists a *principal orbit type* (pot) if there is a dense open subset $\mathcal{P}$ of X such that $G_x \sim G_y$ (conjugate) for all $x,y \in \mathcal{P}$.

THEOREM 5. (Richardson [1]) *Let* G *be a reductive algebraic group,* $G \to GL(V)$ *a rational representation* $(\dim V < \infty)$. *Then there exists a pot for* (G,V).

Proof. (Sketch) The first step is

LEMMA 6. *Under the above assumptions, suppose also there is not an open* G*-orbit. Then there is* $x \in V$, $x \neq 0$ *such that* G_x *is reductive.*

(Note x may not actually be in the pot.)

Now whenever we have an algebraic transformation group $G \times X \to X$, X an affine variety, we can look at $K[X]^G$ = the algebra of G-invariant regular functions on X. This is an affine algebra and so corresponds to an affine variety Y (which can be shown to be identified to the set of closed orbits in X). It is common to write $Y = X/G$.

Next suppose $H \subseteq G$, H reductive, and suppose X is an H-space. Then H acts on the algebraic variety $G \times X$ by $(g,x)\cdot h = (gh, h^{-1}\cdot x)$. Look at the quotient

$$p\colon G\times X \to (G\times X)/H = E.$$

Let $\alpha\colon G \to G/H$ be the canonical projection. Then $G\times X \to G/H$, $(g,x) \to \alpha(g)$ is constant on H-orbits and so induces a map $\pi\colon E \to G/H$. We call E the fiber bundle with principal bundle $\alpha\colon G \to G/H$ and fiber X. If X is linear, $H \to GL(X)$, then E is a homogeneous vector bundle.

Let $F = \pi^{-1}(\alpha(e))$ and define $q\colon X \to F$ by $q(x) = p(e,x)$. q is then an isomorphism of algebraic varieties, F is H-stable, and q is an H-morphism. Now let G act on $G\times X$ by $g\cdot(g_1,x) = (gg_1,x)$. This maps H-orbits to H-orbits and so induces an action of G on E. The morphism $E \to G/H$ is a G-morphism. One checks that each G-orbit on E meets F and that if $a \in F$, $g \in G$, $g\cdot a \in F$, then $g \in H$. Moreover if $\mathcal{U}$ is an H-stable open subset of X, then $p(G\times\mathcal{U})$ is a G-stable open subset of E. We then have

LEMMA 7. (i) *With the above terminology, assume* $\mathcal{U}$ *is an open* H*-stable subset of* X *such that for* $x,y \in \mathcal{U}$, $H_x \sim H_y$ *(resp.* $H_x^0 \sim H_y^0$*). Then* $\mathcal{U}_1 = p(G\times\mathcal{U})$ *is a* G*-stable open subset in* E *and for* $x,y \in \mathcal{U}_1$, $G_x \sim G_y$ *(resp.* $G_x^0 \sim G_y^0$*).*

(ii) *Let* G *act morphically on smooth irreducible algebraic varieties* X,Y. *Let* $\alpha\colon X \to Y$ *be a* G*-morphism such that* $(d\alpha)_x$ *is a linear isomorphism for some* $x \in X$. *Then there exists a principal isotropy subalgebra for* (G,X) $\Leftrightarrow$ *there exists a principal isotropy subalgebra for* (G,Y). *By a principal isotropy subalgebra we mean a Z-open set* $\mathcal{P} \subseteq V$ *such that* $\mathfrak{g}_x \sim \mathfrak{g}_y$ *(or equivalently* $G_x^0 \sim G_y^0$*) for* $x,y \in \mathcal{P}$.

LEMMA 8. *For* $G \to GL(V)$, G *reductive, there is a principal isotropy subalgebra for* (G,V).

Proof. By induction on $(\dim G + \dim V)$. If $\dim G + \dim V = 0$, there is nothing to prove. If G has an open orbit, again we are done. If the fixed point set V^G of G in V is not $\{0\}$, then since G is reductive $V = V^G \oplus V_1$, with V_1 being G-invariant; and we are done by reduction of dimension. So we may assume $V^G = \{0\}$. We may also assume that G is connected.

Now by Lemma 6, there is $x \in V$, $x \neq 0$, such that $G_x = H$ is reductive. We identify the tangent space $T(G\cdot x)_x$ with a vector subspace of V, namely $\mathfrak{g}/\mathfrak{g}_x$. But $T(G\cdot x)_x$ is H-stable and H is reductive; hence there is a complementary subspace W. Let $E = (G\times W)/H$ be the corresponding homogeneous vector bundle. Define a morphism

$$\tau\colon G\times W \to V \qquad \tau(g,w) = g\cdot(x+w).$$

τ is constant on H-orbits $\Rightarrow \tau\colon E \to V$. Moreover τ is a G-morphism. Finally since $V^G = \{0\}$, we have $H = G_x \neq G \Rightarrow (\dim W + \dim H) < (\dim V + \dim G)$. Hence there exists a principal isotropy subalgebra for (H,W) by the induction assumption. Therefore by Lemma 7, there exists a principal isotropy subalgebra for G.

It remains to pass from neutral components to the full stabilizers. That involves some arguments with finite groups whose details we omit (see Richardson [1]).

COROLLARY. *Let the reductive group* G *act morphically on the smooth affine variety* X. *Then there is a* pot *for* (G,X).

Proof. (Sketch) The smoothness essentially allows lifting the picture to the tangent space and thus reducing to Theorem 5.

EXAMPLES. (1) Let $G = SL(n,\mathbb{C})$, $V = \mathbb{C}^n$ and let $G \to GL(V)$ be the natural representation. Then there is a single open orbit, namely $\mathcal{P} = V-\{0\}$, for which the stability group is easily seen to be isomorphic to $SL(n-1,\mathbb{C})\cdot\mathbb{C}^{n-1}$, with $SL(n-1,\mathbb{C})$ acting via the natural representation again.

(2) Let $G = SO(n,\mathbb{C})$, $V = \mathbb{C}^n$ and let $G \to GL(V)$ again be the natural representation. Then the pot is $\mathcal{P} = \{(z_1,\cdots,z_n) \in V: \sum z_i^2 \neq 0\}$. The orbits in $\mathcal{P}$ are the closed hypersurfaces $\sum z_i^2$ = constant, and the stability group of a point in $\mathcal{P}$ is $SO(n-1,\mathbb{C})$.

(3) Let G be a complex semisimple Lie group, $V = \mathfrak{g} = LA(G)$, and let $G \to GL(V)$ be the adjoint representation. Then $\mathcal{P} = \mathfrak{g}'$ is the set of regular elements, the orbits are again closed manifolds, and the stability groups are Cartan subgroups (all of which are conjugate).

Now we want to see what happens when we pass to the rational points.

THEOREM 9. (Lipsman [6]) *Let* $G \to GL(V)$ *be a rational representation* def/k, G *reductive,* $\mathcal{P}$ = pot. *Then the* Z-*open set* $\mathcal{P}_k$ *may be decomposed* $\mathcal{P}_k = \mathcal{P}_1 \cup \cdots \cup \mathcal{P}_n$, *where each* $\mathcal{P}_i$ *is a* p-*adic open set in* V_k *and a* pot *for the action of* G_k *on* V_k. *Furthermore there exists a Borel cross-section* $s_i: \mathcal{P}_i \backslash G_k \to \mathcal{P}_i$ *along which the stability groups are constant.*

Proof. We use the following well-known fact: If $G \times X \to X$ is def/k and $\mathcal{O}$ is a G-orbit in X, then $\mathcal{O} \cap X_k$ decomposes into finitely many G_k-orbits (this is a consequence of the finiteness of the Galois co-homology $H^1(G,\Gamma)$ -- Borel and Serre [1]). In fact we obtain the first part from

LEMMA 10. *Let* G *be reductive and suppose that* $G \times X \to X$ *has only one orbit type, i.e.* $G_x \sim G_y$ $\forall x,y \in X$. *Then there are at most finitely many orbit types for* $G_k \times X_k \to X_k$.

Proof. Let $x \in X_k$. Then $H = G_x$ is def/k, $N = \mathrm{Norm}(H)$ is def/k, and we have bijections

$\beta: G/N \to \mathfrak{C}(H)$ = subgroups of G conjugate to H

$\beta: (G/N)_k \to \mathfrak{C}_k(H)$ = subgroups of G conjugate to H, def /k.

Define the map $\phi: X \to G/N$ by $\phi(z) = \beta^{-1}(G_z)$.

EXERCISE. Check that $\phi|_{X_k} : X_k \to (G/N)_k$. Now G_k acts on $(G/N)_k$ and there are finitely many orbits, say $G_k \cdot g_1 N, \cdots, G_k \cdot g_j N$. Let $X_k^i = \phi^{-1}(G_k \cdot g_i N)$. The proof is completed by showing $z_1, z_2 \in X_k^i \Rightarrow G_{z_1} \sim G_{z_2}$ over k.

We go on with the proof of Theorem 9. For $x \in \mathcal{P}_k$, let W_x be the space of G_x-fixed vectors. Since $x \in \mathcal{P} \cap W_x$, $\mathcal{P} \cap W_x$ is Z-open in W_x. In fact $G \cdot (W_x \cap \mathcal{P}) = \mathcal{P}$. But for $y \in W_x \cap \mathcal{P}$, the stability group $G_y = G_x$. This is because G_y clearly contains G_x, and is conjugate to it as well. Since they are both algebraic groups, they must be the same.

Next take the k-rational points and decompose $\mathcal{P}_k = \mathcal{P}_1 \cup \cdots \cup \mathcal{P}_n$ into finitely many G_k-orbit types. If $x \in \mathcal{P}_i$, then we see that the natural map $G_k \times (W_x)_k \to V_k$ is a submersion (indeed $V_k = (\mathfrak{g}/\mathfrak{g}_x)_k + (W_x)_k$). Therefore by the p-adic implicit function theorem, we have that the $\mathcal{P}_i$ are open sets. The last statement follows from certain arguments involving Effros' results and smoothness of orbits under algebraic group actions.

EXAMPLES. (1) Choose the algebraic structure in Example 2, p. 134 so that taking the $\mathbb{R}$-rational points leads to $G_{\mathbb{R}} \to GL(V_{\mathbb{R}})$ where $G_{\mathbb{R}} = SO(n-1,1)$ and $G_{\mathbb{R}}$ acts on $V_{\mathbb{R}} = \mathbb{R}^n$ as in Example 8, p. 73. From those computations we see that $\mathcal{P}_{\mathbb{R}} = \mathcal{P}_1 \cup \mathcal{P}_2$ corresponding to the different stability groups $SO(n-2,1)$ and $SO(n-1)$.

(2) Let $G = Sp(n,\mathbb{C})$, $V = \mathbb{C}^{2n}$ and $G \to GL(V)$ the natural representation. Then $G_{\mathbb{R}} = Sp(n,\mathbb{R})$, $V_{\mathbb{R}} = \mathbb{R}^{2n}$ and we have

$G_{\mathbb{R}} \to GL(V_{\mathbb{R}})$. In this case $\mathcal{O} = V-\{0\}$, $\mathcal{O}_{\mathbb{R}} = V_{\mathbb{R}}-\{0\}$ and these are open orbits. We leave it to the reader to check that the stability group of a point $x \in \mathcal{O}_{\mathbb{R}}$ is of the form

$$(G_x)_{\mathbb{R}} = Sp(n-1,\mathbb{R}) \cdot \mathcal{H}_{n-1}$$

where $\mathcal{H}_{n-1}$ is the Heisenberg group of dimension $2n-1$ with one-dimensional center, and $Sp(n-1,\mathbb{R})$ acts as in Example 4, p. 81 (case $n = 3$).

The situation is not so transparent with regard to the nature of the stability groups. Nor is the situation of U unipotent particularly well-understood -- neither the orbit types, nor the stabilizers. We simply state some questions that are currently under investigation in this regard.

QUESTIONS. I. U abelian -- or what amounts to the same thing $G \to GL(V)$.

(1) If G is semisimple and there is an open orbit $\mathcal{O}$, then for $x \in \mathcal{O}$ is it true that $R(G_x)$ is unipotent, i.e. G_x is a semidirect product of semisimple and unipotent groups?

(2) If G is simple and the representation is irreducible, then is it true that $\mathcal{O}$ consists either of an open orbit or of only closed orbits? (Note the latter holds $\Longleftrightarrow$ the stability groups in general position are reductive).

(3) If G is semisimple, $x \in \mathcal{O}$, then is $R(G_x)$ nilpotent?

II. U unipotent.

(1) By the Corollary to Richardson's Theorem and the fact that in unipotent groups the generic points in $\hat{U}_k$ form a smooth affine variety, we see that there is a pot for H_k on $\hat{U}_k$. Does there exist a pot for the Mackey obstructions ω_γ? What can be said about the stabilizers?

(2) Is there a Kirillov theory for groups which are semidirect products of semisimple and unipotent groups?

E. CCR PROPERTY FOR ALGEBRAIC GROUPS

The last topic in this chapter is a classification of CCR algebraic groups. We shall make use here of the dual topology on $\hat{G}$ (See Fell [1]). In particular, we shall use the well-known fact that $\hat{G}$ is T_1 if and only if G is CCR (Dixmier [5]).

Let G be a connected algebraic group, $G = HU$ a Levi decomposition. The group H acts on U via inner autormorphism, thus it acts on $\mathfrak{u}$ and $\mathfrak{u}^*$ as well. Set $M = \{h \in H: h\cdot u = u, \ \forall u \in U\}$. It is easily seen that

$$M = \{h \in H: h\cdot W = W, \ \forall W \in \mathfrak{u}\} = \{h \in H: h\cdot \phi = \phi, \ \forall \phi \in \mathfrak{u}^*\}.$$

THEOREM 1. (Lipsman [5]) *If* G_k *is CCR, then* H/M *is anisotropic over* k.

Proof. As usual we write Z for a maximal k-split torus of G in H. What we must show is that if G_k is CCR, then $Z \subseteq M$.

Let $C = Z_U$ and consider the ascending central series $\{e\} \subseteq C = C_0 \subseteq C_1 \subseteq \cdots \subseteq C_q = U$. The proof is by induction on q. Take first the case $q = 0$, that is $U = C$ is abelian. Now Z acts as a group of linear transformations on $\mathfrak{u}^*$ by the co-adjoint representation. This determines a representation

$$\rho: Z \to GL(\mathfrak{u}^*)$$

which is def/k. Let $\Phi = (\Phi_1,\cdots,\Phi_r)$ be a basis for $\mathfrak{u}^*$. Denote by $\lambda_\Phi: GL(\mathfrak{u}^*) \to GL(r,K)$ the corresponding homomorphism. Choose $\Phi_i \in \mathfrak{u}^*_k$, $\lambda_\Phi: GL(\mathfrak{u}^*_k) \to GL(r,k)$. Since Z is a torus there exists $T \in GL(r,K)$ such that $T^{-1}\lambda_\Phi(\rho(Z))T \subseteq D(r,K)$, the diagonal matrices. Moreover Z k-split guarantees that T can be chosen in $GL(r,k)$.

Then setting $\phi = (\lambda_\Phi^{-1}T)(\Phi)$, $\phi = (\phi_1,\cdots,\phi_r)$, $\phi_i \in \mathfrak{n}_k^*$, we have

$$z\cdot\phi_i = \chi_i(z)\phi_i, \quad 1 \leq i \leq r, \quad \chi_i \in X(Z), \quad \chi_i|_{Z_k} : Z_k \to k^*.$$

If $z = (z_1,\cdots,z_m)$ then $\chi_i(z) = z_1^{b_1}\cdots z_m^{b_m}$, $b_i \in \mathbf{Z}$. It follows from the structure of local fields that $\chi_i(Z_k)$ is either a subgroup of finite index in k^* or is trivial. We must show that all χ_i are trivial. If not let $\chi = \chi_1 \neq 1$, $\phi = \phi_1$. Then a simple argument shows it is no loss of generality to assume $\chi(Z_k) = k^*$.

Apply the extension prodedure to $U_k \subseteq G_k$. Fix ε as in Chapter V, section C, and identify $\mathfrak{n}_k^*$ with $\hat{U}_k$ via

$$\psi(\exp X) = \varepsilon(\psi(X)), \quad \psi \in \mathfrak{n}_k^*, \quad X \in \mathfrak{n}_k.$$

For any $x \in k^*$, there is $z \in Z_k$ such that $\chi(z) = x$. Consider the characters $x\phi$, $x \in k^*$. Choose $x \to 0$ in k^*. Then the characters $x\phi \to 1$ in $\hat{U}_k$. The stability groups are all the same since H acts by linear transformations on $\mathfrak{n}$. That is $(H_k)_{x\phi} = (H_k)_\phi = H_1$ say. Form the representations

$$\gamma_x : hu \to \varepsilon(x\phi(\log u)), \quad h \in H_1, \quad u \in U_k.$$

Clearly $\gamma_x \to 1$ in $(H_1U_k)^\wedge$. Then by the Mackey theory (since all $x\phi$ are in the same orbit) the representations $\operatorname{Ind}_{H_1U_k}^{G_k} \gamma_x$ are all irreducible and equivalent. Call the common class $\{\pi\}$. Then by the continuity of induction (Fell [3])

$$\pi \to \operatorname{Ind}_{H_1U_k}^{G_k} 1.$$

The latter is a reducible representation, lying over the trivial orbit in $\hat{U}_k$. Hence it is not a multiple of π. Therefore $\{\pi\}$ is not a closed point of the dual, a contradiction to the fact that G_k is CCR.

Next assume U is unipotent, but not abelian $(q > 0)$. Take $C = Z_U$. C is normal in G and G/C is a connected linear algebraic group with Levi decomposition $G/C = H\cdot U/C$. Moreover in this case $(G/C)_k = G_k/C_k$ and since the CCR property is preserved by quotients G_k/C_k is CCR. Therefore by the induction hypothesis

$$Z \subseteq M_C = \{h \in H: h\cdot u \equiv u \bmod C, \quad \forall u \in U\}$$

$$= \{h \in H: \forall X \in \mathfrak{n}, \exists W_{X,h} \in \mathfrak{c} = LA(C) \text{ such that } h\cdot X = X+W_{X,h}\}.$$

In particular

$$z\cdot X = X+W_{X,z}, \quad z \in Z, \quad X \in \mathfrak{n}, \quad W_{X,z} \in \mathfrak{c}.$$

Next we work as before, this time decomposing the action of Z on $\mathfrak{n}$ (instead of $\mathfrak{n}^*$). Without working out the details, we mention that using the above equation and the assumption that $z\cdot u_1 = \chi_1(z)u_1$, $\chi_1 \not\equiv 1$, it is possible to define a linear function $\phi \in \mathfrak{n}_k^*$ for which $\mathfrak{n}_k$ itself is a maximal subordinate subalgebra, i.e. ϕ lifts to a unitary character in $\hat{U}_k$. The same type of reasoning as in the previous case (using the dual topology) then contradicts CCR. See Lipsman [5] for the details.

We finish the section with some observations.

(1) The converse of Theorem 1 would be true if we knew that semisimple p-adic groups were CCR.

(2) Since whenever a reductive group H is anisotropic over k the group of rational points H_k is compact, it follows easily that: if G_k is CCR, then G_k is unimodular. It has been conjectured that this is always true, that is any CCR group must be unimodular. Whether it is true or not is still an open problem.

(3) If G is def/$\mathbb{R}$ and $G_{\mathbb{R}}$ is CCR, then by structure theory of real Lie and algebraic groups, it can be seen that $G_{\mathbb{R}}^{\circ} = H_1S_1U_1$, where U_1 is a normal simply connected nilpotent Lie group, H_1 and S_1 are connected reductive Lie groups, S_1 is

compact, $H_1S_1 \cap U_1 = \{e\}$, $H_1 \cap S_1$ is finite, and H_1 commutes with S_1U_1. One wonders if there is some kind of structure theorem for general (not algebraic) CCR real Lie groups.

(4) Finally there are a few more outstanding conjectures (due to several people) about possible necessary and sufficient conditions for CCR. Specifically let G be a Lie group. The following are all easily seen to be sufficient, but it is unknown whether any are necessary:

(a) G is traceable.

(b) There exists an invertible operator $\Delta \in \mathfrak{U}(\mathfrak{g})$ such that $\pi(\Delta^{-1})$ is trace class for every $\pi \in \hat{G}$.

(c) There exists $n > 0$ such that $\pi(C_0^{(n)}(G)) \subseteq$ Hilbert-Schmidt operators for every $\pi \in \hat{G}$; here $C_0^{(n)}(G))$ denotes the n-times continuously differentiable functions of compact support on G.

CHAPTER VI. SOLVABLE GROUPS

In this chapter, I give a very short introduction to the representation theory of solvable groups. For more detail in the real case the reader is referred to the excellent notes of S. Quint [1]. The material in the p-adic case is very recent and has not really been digested by the mathematical community yet.

A. LIE GROUPS

The simplest class of solvable Lie groups beyond nilpotent is the collection of exponential groups.

DEFINITION. A Lie group G is called *exponential* if $\exp: \mathfrak{g} \to G$ is a diffeomorphism.

In such a case G must be solvable and simply connected. For those groups we have the following result of Bernat [1].

THEOREM 1. *For each* $f \in \mathfrak{g}^*$, *there exists a subordinate subalgebra* $\mathfrak{h} \subseteq \mathfrak{g}$ *such that* $\rho(f,\mathfrak{h}) = \mathrm{Ind}_{\exp \mathfrak{h}}^{G} \chi_f$, $\chi_f(\exp X) = e^{if(X)}$, $X \in \mathfrak{h}$, *is irreducible. Such algebras have maximal dimension among subordinate subalgebras. Moreover* $\rho(f,\mathfrak{h})$ *depends only on* f, *say* $\rho(f)$, *and further depends only on the* G-*orbit of* $f \in \mathfrak{g}^*$. *Finally* $G\cdot f \to \rho(f)$ *is a bijection of* $\mathfrak{g}^*/G$ *onto* $\hat{G}$.

The only difference from the nilpotent case is that the maximal subordinate subalgebra condition is necessary but not sufficient for the irreducibility of the representation. Pukanszky [2] found the precise necessary and sufficient condition.

THEOREM 2. *The representation* $\rho(f,\mathfrak{h})$ *is irreducible if and only if* $\mathfrak{h}$ *is of maximal dimension among subordinate subalgebras and the orbit* $\mathcal{O}$ *of* f *in* $\mathfrak{g}^*$ *contains the affine space* $f+\mathfrak{h}^{\perp}$, $\mathfrak{h}^{\perp} = \{\phi \in \mathfrak{g}^*: \phi(\mathfrak{h}) = 0\}$.

The maximality condition is as usual equivalent to $\dim \mathfrak{h} = \dim \mathfrak{g} - \frac{1}{2}\dim \mathcal{O}$. $\mathcal{O} \supseteq f+\mathfrak{h}^{\perp}$ is called the *Pukanszky condition.* It is known to be equivalent to $\exp \mathfrak{h}\cdot f = f+\mathfrak{h}^{\perp}$. It is automatic for $\mathfrak{h}$'s of maximal dimension when $\mathfrak{g}$ is nilpotent.

EXAMPLE. Let $G = \{\begin{pmatrix} a & b \\ 0 & a^{-1}\end{pmatrix} : a > 0,\ b \in \mathbb{R}\}$; the so-called "ax+b group". The Lie algebra is $\mathfrak{g} = \{\begin{pmatrix} \alpha & \beta \\ 0 & -\alpha\end{pmatrix} : \alpha,\beta \in \mathbb{R}\}$ and

$$\exp\begin{pmatrix} \alpha & \beta \\ 0 & -\alpha\end{pmatrix} = \begin{pmatrix} e^{\alpha} & \beta\frac{\sin h\alpha}{\alpha} \\ 0 & e^{-\alpha}\end{pmatrix}.$$

It follows easily that G is exponential solvable. Let us compute the adjoint and co-adjoint representations:

$$g = \begin{pmatrix} a & b \\ 0 & a^{-1}\end{pmatrix} \in G,\quad X = \begin{pmatrix} \alpha & \beta \\ 0 & -\alpha\end{pmatrix} \in \mathfrak{g},\quad f(X) = u\alpha+v\beta,$$

$$f \in \mathfrak{g}^*,\quad u,v \in \mathbb{R}$$

$$\mathrm{Ad}\, g(X) = gXg^{-1} = \begin{pmatrix} \alpha & a^2\beta-2ab\alpha \\ 0 & -\alpha\end{pmatrix}$$

$$\mathrm{Ad}^*g^{-1}(f)(X) = f(\mathrm{Ad}g(X)) = u\alpha + v(a^2\beta-2ab\alpha)$$

$$= (u-2abv)\alpha + (a^2v)\beta.$$

Hence there are two types of orbits:

(a) $f = (u,0)$. These orbits are points. They give rise to the characters of G which are trivial on the normal subgroup $\{\begin{pmatrix} 1 & b \\ 0 & 1\end{pmatrix} : b \in \mathbb{R}\}$.

(b) $\mathcal{O}^+ = \{f = (u,v) : v > 0\}$, $\mathcal{O}^- = \{f = (u,v)\ v < 0\}$. A maximal subordinate subalgebra in either case would be $\mathfrak{h} = \{\begin{pmatrix} 0 & \beta \\ 0 & 0 \end{pmatrix} : \beta \in R\}$. $\mathfrak{h}$ satisfies the Pukanszky condition both times. Thus we obtain two infinite-dimensional irreducible representations π^+, π^-. Note that the results here are in agreement with those obtained by the Mackey theory (See Chapter III, section A, Example 1). In principal every solvable group is built up of semidirect products with abelian groups and so could be handled by the group extension procedure. In practice this is exceedingly tedious -- what's more, no uniform description of $\hat{G}$ has ever been found in this way. Only generalizations of the Kirillov orbit thoery have yielded such a parameterization.

Finally note that $\mathfrak{h}_m = \{\begin{pmatrix} m\alpha & \alpha \\ 0 & -m\alpha \end{pmatrix} : \alpha \in \mathbb{R}\}$ is also a maximal subordinate subalgebra. However for $f = (0,1)$, $\rho(f,\mathfrak{h}_m)$, $m \neq 0$, is not irreducible (since $\mathfrak{h}_m$ does not satisfy the Pukanszky condition). In fact $\rho(f,\mathfrak{h}_m) = \pi^+ \oplus \pi^-$. That behavior is a special case of a general result due to Vergne [1].

THEOREM 3. *Let* $f \in \mathfrak{g}^*$, $\mathfrak{h}$ *a subordinate subalgebra of maximal dimension. Then the affine space* $f+\mathfrak{h}^\perp$ *is, up to a closed set of lower dimension, the intersection of* $f+\mathfrak{h}^\perp$ *with a finite number of orbits* $\mathcal{O}_1,\cdots,\mathcal{O}_k$; *each* $\mathcal{O}_i \cap (f+\mathfrak{h}^\perp)$ *being open in* $f+\mathfrak{h}^\perp$ *and having a finite number* n_i *of connected components. Then* $\rho(f,\mathfrak{h}) = \sum^{\oplus} n_i\rho_i$, $\rho_i = \rho(\mathcal{O}_i)$.

REMARKS. (1) The thesis of S. Quint [2] is involved with dropping the maximality assumption on $\mathfrak{h}$.

(2) Exponential solvable groups are not unimodular in general, and so in line with previously expressed philosophy it is unlikely that they will be traceable. (It is an interesting exercise to show that the ax+b group is not traceable.) However, certain "characters" of these groups have been computed. Roughly speaking, for C^∞

functions ϕ supported on a small enough neighborhood of the identity, $\pi(\phi)$ can be shown to be trace class for $\pi \in \hat{G}$, and one can compute $\mathrm{Tr}\ \pi(\phi)$. For this information (due to Pukanszky and Duflo) see Bernat et al [1].

Now we wish to pass to more general solvable groups. In the non-exponential case, examples show that: on the one hand, it may not be possible to find real polarizations; and on the other hand, even if you can find them, the corresponding induced representations may be reducible -- even when the Pukanszky condition is satisfied. These examples may be found in Quint [1]. The former occurs on the oscillator group, the latter is evident already in the universal covering group of the Euclidean motion group of the plane. To compensate for these deficiencies one must consider complex polarizations and holomorphic induction. The general theory is due to Auslander and Kostant [1], based on prior work of Streater [1], Bargmann [1] and others. We outline the theory.

Let G be a simply connected solvable Lie group, $\mathfrak{g} = LA(G)$, $f \in \mathfrak{g}^*$. As usual we write B_f for the skew-symmetric bilinear form $B_f(x,y) = f[x,y]$, $x,y \in \mathfrak{g}$. Also set $\mathcal{O} = G\cdot f \subseteq \mathfrak{g}^*$, G_f = the stability group, and $\mathfrak{g}_f = LA(G_f) = \{x \in \mathfrak{g}:\ B_f(x,y) = 0 \ \ \forall y \in \mathfrak{g}\}$. B_f is non-degenerate on $\mathfrak{g}/\mathfrak{g}_f$; so $\dim(\mathfrak{g}/\mathfrak{g}_f)$ is even. After identifying $\mathfrak{g}/\mathfrak{g}_f$ to the tangent space to $\mathcal{O}$ at f, we denote the image of B_f on that tangent space by ω_f. Then $f \to \omega_f$ is a differential 2-form on $\mathcal{O}$. The form ω is closed and G-invariant (i.e. $\mathcal{O}$ is a homogeneous symplectic manifold). If $\dim \mathcal{O} = 2d$, then ω^d is a volume form on $\mathcal{O}$.

Write $n = \dim \mathfrak{g}$, $m = \dim \mathfrak{g}_f$, $n-m = 2d$. We write B_f for the extension of B_f to the complexification of $\mathfrak{g}_c$. Let V be a maximal isotropic complex subspace, $\mathfrak{g}_f \subseteq V$ and $\dim_{\mathbb{C}} V = m+d$. Consider such subspaces $\mathfrak{h}$ which are in addition subalgebras. Then

$\mathfrak{h} \cap \overline{\mathfrak{h}}$ is a subalgebra of $\mathfrak{g}_c$, invariant under $x \to \overline{x}$, and so $\mathfrak{h} \cap \mathfrak{g} = \mathfrak{d}$ is a real subalgebra such that $\mathfrak{h} \cap \overline{\mathfrak{h}} = \mathfrak{d}_c$ is its complexification. Also $\mathfrak{h} + \overline{\mathfrak{h}}$ is a subspace of $\mathfrak{g}_c$ def/$\mathbb{R}$, but not a subalgebra in general. Let $\mathfrak{e}_c = \mathfrak{h} + \overline{\mathfrak{h}}$.

DEFINITION. $\mathfrak{h} \subseteq \mathfrak{g}_c$ is called a *polarization* for f if

(1) $\mathfrak{h}$ is maximal totally isotropic for B_f.

(2) $\mathfrak{h} + \overline{\mathfrak{h}} = \mathfrak{e}_c$ is a subalgebra.

If $\mathfrak{h}$ is a polarization, then $(\mathfrak{h} + \overline{\mathfrak{h}}) \cap \mathfrak{g} = \mathfrak{e}$ is a real subalgebra with $\mathfrak{e}_c$ as its complexification. $\mathfrak{h}$ is called *real* if $\mathfrak{h} = \overline{\mathfrak{h}}$.

From the definitions of $\mathfrak{e}_c$ and $\mathfrak{d}_c$ and the fact that $\mathfrak{h}$ is a maximal totally isotropic subspace for B_f, it follows that $\mathfrak{d}_c$ is the orthogonal complement of $\mathfrak{e}_c$ with respect to B_f. Hence $\mathfrak{d}$ is the orthogonal complement of $\mathfrak{e}$ for the real form B_f. Thus $\mathfrak{e}/\mathfrak{d}$ carries a non-degenerate skew-symmetric form (also denoted B_f). In particular $\mathfrak{e}/\mathfrak{d}$ is even-dimensional.

Next note that

$$(\mathfrak{e}/\mathfrak{d})_c \cong \frac{\mathfrak{e}_c}{\mathfrak{d}_c} = \frac{\mathfrak{h} + \overline{\mathfrak{h}}}{\mathfrak{h} \cap \overline{\mathfrak{h}}} = \frac{\mathfrak{h}}{\mathfrak{h} \cap \overline{\mathfrak{h}}} \oplus \frac{\overline{\mathfrak{h}}}{\mathfrak{h} \cap \overline{\mathfrak{h}}} = \frac{\mathfrak{h}}{\mathfrak{d}_c} \oplus \frac{\overline{\mathfrak{h}}}{\mathfrak{d}_c}.$$

Define $J\colon (\mathfrak{e}/\mathfrak{d})_c \to (\mathfrak{e}/\mathfrak{d})_c$ to be multiplication by $\begin{cases} +i & \text{on } \mathfrak{h}/\mathfrak{d}_c \\ -i & \text{on } \overline{\mathfrak{h}}/\mathfrak{d}_c \end{cases}$. J leaves $\mathfrak{e}/\mathfrak{d}$ invariant and so defines a complex structure thereon.

EXERCISES. (1) Show that $\mathfrak{e}/\mathfrak{d}$ is isomorphic with its complex structure to $\mathfrak{h}/\mathfrak{d}_c$.

(2) Show that $B_f(x,y) = B_f(Jx,Jy)$ on $\mathfrak{e}/\mathfrak{d}$.

Next make the definitions

$$S_f(x,y) = B_f(x,Jy) \qquad H_f = S_f + iB_f.$$

H_f is a non-degenerate hermitian form on $\mathfrak{e}/\mathfrak{d}$. It also can be considered to be a hermitian form on $\mathfrak{h}$ with null space $\mathfrak{d}_c$ which satisfies $H_f(x,y) = 2iB_f(x,\bar{y})$, $x,y \in \mathfrak{h}$.

DEFINITION. $\mathfrak{h}$ is called *positive* if H_f is positive-definite, i.e. if $H_f(x,x) \geqq 0$, $x \in \mathfrak{e}/\mathfrak{d}$ or $2iB_f(x,\bar{x}) \geqq 0$, $x \in \mathfrak{h}$.

Note of course that real polarizations are positive. Also since $\mathfrak{g}_f \subseteqq \mathfrak{h}$, $\mathfrak{h}$ is G_f^0-invariant. G_f-invariance of $\mathfrak{h}$ is not automatic however.

LEMMA 4. *If* $\mathfrak{h}$ *is stable under* G_f, *then* $D = (\exp \mathfrak{d})G_f$ *is closed and has* $\exp \mathfrak{d}$ *for its neutral component.*

If $\mathfrak{h}$ is G_f-invariant, we say that is satisfies the *Pukanszky condition* if $f+\mathfrak{e}^+ \subseteqq \mathcal{O}$. We say that f is *integral* if there is a character σ_f of G_f such that $\sigma_f(\exp x) = e^{2\pi i f(x)}$, $x \in \mathfrak{g}_f$. Note that σ_f always exists on G_f^0; extending it to G_f may or may not be possible. If it is possible it may be done in exactly this many ways:

For σ_f fixed, any other extension is given by

$$\sigma_f'(g) = \sigma_f(g)s(g), \qquad s \in (G_f/G_f^0)\hat{} = \pi_1(\mathcal{O})\hat{}.$$

Hence there are $\pi_1(\mathcal{O})\hat{}$ different extensions if any. σ_f' extends uniquely to a character λ_f' of $D = D^0G_f$ such that $\lambda_f'(\exp x) = e^{2\pi i f(x)}$, $x \in \mathfrak{d}$.

LEMMA 5. *Let* $\mathfrak{h}$ *be an invariant polarization satisfying the Pukanszky condition and* admissible *for the nilradical* $\mathfrak{n}$ *of* $\mathfrak{g}$ *(that meaning* $\mathfrak{h} \cap \mathfrak{n}$ *is a polarization for* $f|_{\mathfrak{n}}$*). Then the group* $E = (\exp \mathfrak{e})G_f$ *is also closed.*

Before we describe the representations corresponding to these objects we state

THEOREM 6. *Let* G_f, $G_{f'}$ *be the stabilizers of* $f \in \mathfrak{g}^*$, $f' = f|\mathfrak{n}$. *Then there exists a positive polarization* $\mathfrak{h}$ *satisfying the Pukanszky condition, admissible for* $\mathfrak{n}$ *and such that* $\mathfrak{h}$ *(resp.* $\mathfrak{h} \cap \mathfrak{n}$*) is invariant under* G_f *(resp.* $G_{f'}$*).*

Now take $f \in \mathfrak{g}^*$ integral, and $\mathfrak{h}$ as in Theorem 6. Choose $\sigma_f \in \hat{G}_f$, and $\lambda_f \in \hat{D}$ extending it. Set $m(d) = \Delta_D(d)/\Delta_G(d)$. Note that $m(\exp x) = \exp \operatorname{tr}(\operatorname{ad}_{\mathfrak{g}/\mathfrak{d}}(x))$, $x \in \mathfrak{d}$. Consider the space

$$\mathcal{V} = \{\phi : G \to \mathbb{C} : \phi \text{ is } C^\infty \text{ and}$$

$$(1) \quad \phi(gd) = \lambda_f(d)^{-1} m(d)^{\frac{1}{2}} \phi(g), \quad d \in D, \quad g \in G$$

$$(2) \quad \phi * X = -2\pi i f(X) + \frac{1}{2} \operatorname{tr}(\operatorname{ad}_{\mathfrak{g}/\mathfrak{d}}(X)), \quad X \in \mathfrak{h}$$

$$(3) \quad \int_{G/D} |\phi|^2 < \infty\}.$$

Let $\mathcal{H}(f,\sigma_f,\mathfrak{h},G)$ denote the completion of $\mathcal{V}$ and $\rho(f,\sigma_f,\mathfrak{h},G)$ the representation of G which acts on this space by left translation. Roughly speaking we are inducing from D to G, but between D and E we only take the holomorphic functions.

THEOREM 7. (Auslander and Kostant [1]) *The representation* $\rho(f,\sigma_f,\mathfrak{h},G)$ *is irreducible and independent of the polarization* $\mathfrak{h}$.

This is the key theorem in the theory. It is a very deep result, requiring the full power of the Mackey group extension method for its proof.

THEOREM 8. (Auslander and Kostant [1]) $\rho(f,\sigma_f) \cong \rho(f',\sigma_{f'})$ *if and only if there exists* $g \in G$ *such that* $g \cdot f = f'$ *and* $g \cdot \sigma_f = \sigma_{f'}$.

Thus to each integral orbit, we associate the space $\pi_1(\mathcal{O})^\wedge$, and write $\mathcal{L} = \cup\{\pi_1(\mathcal{O})^\wedge : \mathcal{O} \in \mathfrak{g}^*/G,\ \mathcal{O} \text{ integral}\}$. Then we have a map $\mathcal{L} \to \hat{G}$ which is one-to-one.

THEOREM 9. (Auslander and Kostant [1]) *The group* G *is type I* *if and only if every orbit is* (i) *integral and* (ii) *locally closed*

The orbits are all locally closed if and only if $\mathfrak{g}^*/G$ is countably separated. In such a case then the Kirillov type map $\mathcal{L} \to \hat{G}$ is a bijection.

We leave it to the reader to consult Quint [1] for examples. Also for what information on characters that's available the reader should see Bernat et al [1]. Finally we remark that in case G is unimodular and type I, very little is known about the Plancherel measure of G.

B. p-ADIC GROUPS

The last topic we treat is that of solvable algebraic groups. The material here is taken from Howe's recent preprint [3]. To a considerable extent, it is based on Howe's Kirillov type theory for compact p-adic groups (see Howe [2]), but we don't go into that here. We shall be content with a statement of the main results.

Let k be a non-archimedean locally compact field of characteristic zero, i.e. a p-adic field. Let S be a connected solvable algebraic group def/k, S_k = the group of rational points. Then $S = TU$, a semidirect product of toral and unipotent groups, and $S_k = T_k U_k$. One method of attack would be (as we have done previously) to apply the group extension procedure to $U_k \subseteqq S_k$. Howe obtains one very nice result in that direction, namely

PROPOSITION 1. *Let* $\sigma \in \hat{U}_k$, $(S_k)_\sigma$ = *the stability group,* $(S_k)_\sigma = (T_k)_\sigma U_k$. *Then the obstruction to extending* σ *to a representation of* $(S_k)_\sigma$ *is trivial.*

Another somewhat more novel technique is the following. Howe finds in T_k a certain nice compact group C which obeys his Kirillov theory for compact groups. More generally then, the group CU_k has a very nice Kirillov theory, while the extension T_k/C is discrete abelian and handled fairly easily.

We summarize his main results in the following

THEOREM 2. *Let* $r: \hat{S}_k \to \hat{U}_k/S_k$ *and* $r_1: \hat{S}_k \to \hat{C}_\infty/S_k (C_\infty = CU_k)$ *be the maps derived from the Mackey theory. Let* $\rho \in \hat{S}_k$, $\rho_1 \in \hat{C}_\infty$, $\sigma \in \hat{U}_k$. *Write* $r(\rho) = \mathcal{O}$, $r_1(\rho) = \mathcal{O}_1$. *Suppose* $\rho_1 \in \mathcal{O}_1$ *and let* $\tilde{\mathcal{O}}_1 \subseteq \mathfrak{c}_\infty^*$ *be the corresponding orbit of linear functionals. Let* S_σ *be the stability group of* σ *in* S_k. *Then*

(i) S_k *is type I.*

(ii) S_k *is CCR if and only of the split part of* T *is central in* S *(this is a special case of Chapter V, Theorem E1).*

(iii) *The Mackey obstruction to extending* σ *from* U_k *to* S_σ *is trivial (Proposition 1).*

(iv) *All irreducible representations* ρ *of* S_k *are induced from finite-dimensional representations of suitable subgroups. All irreducible representations* ρ_1 *of* C_∞ *are induced from one-dimensional representations of suitable subgroups.*

(v) *The restriction of* $\rho \in \hat{S}_k$ *to* U_k *is multiplicity-free. The restriction of* ρ *to* C_∞ *is discretely decomposable and multiplicity-free. It is the direct sum of representations* $s \cdot \rho_1$, $s \in S_k/C_\infty$, *each appearing once.*

(vi) ρ *is CCR if and only if* $\tilde{\mathcal{O}}_1$ *is closed in* $\mathfrak{c}_\infty^*$.

(vii) *Any representation* $\rho' \in r^{-1}(\mathcal{O})$ *is of the form* $\tau \otimes \rho$ *where* τ *is a character of* S_k/U_k. *Moreover* $\tau \otimes \rho$ *is equivalent to* ρ *if and only if* τ *is trivial on* S_ψ/U_ψ, ψ = *the linear functional corresponding to* $\sigma \in \mathcal{O}$. *Thus* $r^{-1}(\mathcal{O}) \cong \hat{T}_k/T_\sigma^\perp \cong \hat{T}_\sigma$.

For further details on these and related results, the reader is referred to Howe [3].

APPENDIX

A. INDUCED REPRESENTATIONS

It is our purpose here to provide a brief summary of the main ideas and results on induced representations which are fundamental to the subject matter of this entire treatise. The idea of forming induced representations -- that is, of inducing a representation of a group G from a representation of a subgroup H -- goes back to Frobenius in the last century. His work of course was in the case G finite. The possibility of forming induced representations for infinite groups was later explored (usually in primitive or ad hoc fashion) by many authors, most notably by the Russians Gelfand, Namiark, Graev, etc. The subject was finally givena firm footing by Mackey in the 1940's and early 1950's. It was a fundamental achievement, without which the theory of group representations might still be crawling around on its hands and knees.

We begin with the notion of a quasi-invariant measure. Let G be a locally compact group and suppose X is a right Borel G-space. That is, there is a Borel map $X \times G \to X$, $(x,g) \to x \cdot g$, such that:

$$x \cdot e = x, \quad \forall x \in X, \quad \text{and} \quad x \cdot (g_1 g_2) = (x \cdot g_1) \cdot g_2$$

or said otherwise $g \to (g\colon x \to x \cdot g)$ is an anti-homomorphism of G into the group of Borel automorphisms of X. Next suppose that X carries a σ finite positive Borel measure μ. We say that μ is *quasi-invariant* under the action of G if for every $g \in G$, the measures μ and $\mu \cdot g$ are equivalent. By $\mu \cdot g$ we mean the measure that assigns to a Borel set $E \subseteq X$ the value $(\mu \cdot g)(E) = \mu(E \cdot g)$, $E \cdot g = \{x \cdot g\colon x \in E\}$. If μ is quasi-invariant, then there is a non-negative Borel function $\alpha(x,g)$ such that

$$\int_X f(x)d\mu(x) = \int_X f(x\cdot g)\alpha(x,g)d\mu(x), \qquad f \in L_1(X,\mu).$$

EXERCISE. Check that α must satisfy

$$\alpha(x,g_1g_2) = \alpha(x,g_1)\alpha(x\cdot g_1,g_2).$$

In such a case we can define a unitary representation U of G as follows: the Hilbert space is $L_2(X,\mu)$ and G acts via

$$U_gf(x) = f(x\cdot g)\alpha(x,g)^{\frac{1}{2}}, \qquad f \in L_2(X,\mu).$$

We leave it to the reader as a simple exercise to check that this defines a continuous unitary representation of G.

We now specialize the space X somewhat (in essence, we take the case G transitive in the preceding). Let H be a closed subgroup of the locally compact group G. The space $X = G/H$ of right cosets is then a right Borel G-space, $(Hg,g_1) \to Hgg_1$. We often write $\bar{x} = Hx$, $x \in G$. The point is that G/H carries a quasi-invariant measure. To see that we begin with a lemma.

LEMMA 1. *There exists a strictly positive continuous function* q *on* G *such that* $q(e) = 1$, $q(hx) = \Delta_H(h)\Delta_G(h)^{-1}q(x)$, $h \in H$, $x \in G$.

Proof. It is possible to find a Borel cross-section $s\colon G/H \to G$, $s(\bar{e}) = e$, and s takes compact sets of G/H to relatively compact sets in G (see Mackey [3]). Then define $q(hs(\bar{x})) = \Delta_H(h)\Delta_G(h)^{-1}$, $h \in H$, $\bar{x} \in G/H$. Since every $x \in G$ is written uniquely $x = hs(\bar{x})$, $h \in H$, we see that q is well-defined and Borel. To see that q can actually be chosen continuous see e.g. Bruhat [1].

Now for $f \in C_0(G)$, set $f'(\bar{x}) = \int_H f(hx)dh$, dh = right Haar measure.

EXERCISE. Show that the map $f \to f'$, $C_0(G) \to C_0(G/H)$ is surjective.

Then define a measure (which we write $\overline{dg}$) on G/H by the formula

$$\int_{G/H} f'(\overline{g})d\overline{g} = \int_G f(x)\,q(x)dx, \quad dx = \text{right Haar measure.}$$

Of course to justify that $d\overline{g}$ is well-defined one needs to show that: $f \in C_0(G)$, $\int_H f(hx)dh = 0$, $\forall x \in G \Rightarrow \int_G f(x)q(x)dx = 0$ -- for that, see Mackey [3].

THEOREM 2. $d\overline{g}$ *is quasi-invariant under* G. *Moreover any other quasi-invariant measure on* G/H *is equivalent to* $d\overline{g}$. *Finally we have the equation*

$$\int_{G/H} f(\overline{x})d\overline{x} = \int_{G/H} f(\overline{x}\cdot g)\left[\frac{q(xg)}{q(x)}\right] d\overline{x}, \qquad f \in C_0(G/H).$$

Proof. The equation of the theorem is easily derived from the equation which defines the measure $d\overline{x}$. The first statement follows immediately. The justification of the second statement may be found in Mackey [3].

EXERCISE. Show that G/H has an invariant measure if and only if $\Delta_G|_H = \Delta_H$.

In light of the previous results, it follows that the formula

$$U_g f(\overline{x}) = f(\overline{x}\cdot g)\left[\frac{q(xg)}{q(x)}\right]^{\frac{1}{2}} d\overline{x}, \qquad f \in L_2(G/H)$$

defines a unitary representation of G. In fact this is the representation of G induced by the trivial representation of H. However we want to induce representations other than the trivial one. For that it is customary and convenient to realize the Hilbert spaces as function spaces on G rather than on G/H.

DEFINITION. Let γ be a representation of H on $\mathcal{H}_\gamma$. Consider the space $\mathcal{H}(\gamma)$ of all functions from G to $\mathcal{H}_\gamma$ satisfying

(1) $x \to (f(x),\xi)$ is Borel $\forall \xi \in \mathcal{H}_\gamma$,

(2) $f(hx) = \gamma(h)f(x)$, $h \in H$, a.a. $x \in G$

(3) $\int_{G/H} \|f(\bar{x})\|^2 d\bar{x} < \infty$,

where we identify functions which differ on sets of measure zero. We set

$$\pi(g)f(x) = f(xg)\left[\frac{q(xg)}{q(x)}\right]^{\frac{1}{2}}, \qquad f \in \mathcal{H}(\gamma),$$

and call π the representation of G *induced* from H by γ. We write $\pi = \mathrm{Ind}_H^G \gamma$.

EXERCISES. (1) Check that π is a continuous unitary representation of G.

(2) In case $\gamma = 1_H$, show that the map $\mathcal{H}(\gamma) \to L_2(G/H)$, $f \to F$, $F(\bar{x}) = f(s(\bar{x}))$, $\bar{x} \in G/H$, sets up an equivalence between $\mathrm{Ind}_H^G 1_H$ and the representation U defined previously. If $H = \{e\}$, the resulting representation is the right regular representation of G.

(3) If we begin the procedure with another function q' satisfying the conditions of Lemma 1 and call the resulting representation π', show that $\pi \cong \pi'$.

(4) Let $D = \{\phi \in C_0^+(G): \int_H \phi(hx)dh \leqq 1\}$ and set $\mathcal{H}(\gamma)'$ to be the space of functions from G to $\mathcal{H}_\gamma$ satisfying

(a) $x \to (f(x),\xi)$ is Borel $\forall \xi \in \mathcal{H}_\gamma$,

(b) $f(hx) = q^{\frac{1}{2}}(h)\,\gamma(h)f(x)$, $h \in H$, a.a. $x \in G$

(c) $\sup_{\phi \in D} \int_G \|f(x)\|^2\,\phi(x)\,dx < \infty$,

identifying functions equal almost everywhere as usual. Show that the unitary map

$$\mathcal{H}(\gamma) \to \mathcal{H}(\gamma)'$$

$$f \to F, \quad F(hs(\bar{x})) = q^{\frac{1}{2}}(h)F(hs(\bar{x})), \quad h \in H, \quad \bar{x} \in G/H$$

converts $\pi = \mathrm{Ind}_H^G \gamma$ into an equivalent representation π' on $\mathcal{H}(\gamma)'$ which acts via

$$\pi'(g)F(x) = F(xg), \qquad F \in \mathcal{H}(\gamma)'.$$

(5) Reformulate the definition of an induced representation so as to be phrased in terms of left actions.

We conclude this section by stating some of the elementary properties of these induced representations. Although elementary, the actual proofs often involve tedious measure-theoretic considerations (see Mackey [3]). The notation is $\pi = \mathrm{Ind}_H^G \gamma$.

(1) π in general fails to be irreducible, even when γ is. However if π is irreducible, then γ must also be irreducible (see number (5)).

(2) If T is an automorphism of G whose restriction to H is an automorphism of H, then $T\pi \cong \mathrm{Ind}_H^G T\gamma$. By $T\pi$, we mean the representation $T\pi(x) = \pi(T^{-1}x)$, $x \in G$.

(3) *Induction in Stages.* If $K \subseteq H \subseteq G$ are closed subgroups and σ is a unitary representation of K, then

$$\mathrm{Ind}_K^G \sigma \cong \mathrm{Ind}_H^G(\mathrm{Ind}_K^H \sigma).$$

Taking $K = \{e\}$, σ trivial, we see that inducing the regular representation of H to G yields the regular representation of G.

(4) If A,B are closed subgroups of C,D respectively, and π_1,π_2 are representations of A,B respectively, then

$$\mathrm{Ind}_A^C \pi_1 \times \mathrm{Ind}_B^D \pi_2 \cong \mathrm{Ind}_{A\times B}^{C\times D} \pi_1 \times \pi_2.$$

In particular if $G = HN$ is a direct product of subgroups and $K \subseteq H$, then for representations σ of K we have

$$\mathrm{Ind}_K^G \sigma \cong \mathrm{Ind}_K^H \sigma \times \lambda_N,$$

λ_N = the regular representation of N.

(5) If the unitary representation γ of H is a direct integral of representations ω

$$\gamma = \int_\Omega^\oplus \omega \, d\nu(\omega)$$

then

$$\operatorname{Ind}_H^G \gamma \cong \int_\Omega^\oplus \operatorname{Ind}_H^G \omega \, d\nu(\omega).$$

Finally we note that a multitude of examples of induced representations occur throughout the entire body of these notes.

B. THE IMPRIMITIVITY THEOREM

The definition of a system of imprimitivity and the imprimitivity theorem that we state here are not nearly as general as is possible. However, they will suffice for our purposes.

Let π be a unitary representation of the locally compact group G. Suppose also that we have a right Borel G-space X on which there is a projection-valued measure P taking values in the space of π. We say that P is a *system of imprimitivity* for π (based on X) if

$$\pi(x)P_E\pi(x)^{-1} = P_{E\cdot x^{-1}}, \quad x \in G, \quad E \in \mathcal{B}(X).$$

EXAMPLE. Let $H \subseteqq G$ be a closed subgroup and γ a unitary representation of H. Set $\pi = \operatorname{Ind}_H^G \gamma$ and take $X = G/H$. G acts in the usual way. Then for Borel sets $E \subseteqq X$, set

$$(P_E^\gamma f)(x) = \chi_E(\bar{x})f(x), \quad f \in \mathcal{H}_\pi.$$

The simple calculation verifying that P^γ is a system of imprimitivity for π (based on G/H) is left to the reader.

In the preceding example, the Borel space on which the system of imprimitivity is based is a homogeneous space of G. Such a system is quite naturally called a *transitive* system of imprimitivity. The content of the imprimitivity theorem is that this example is the most general type of transitive system of imprimitivity.

THEOREM 1. (Mackey [1]) *Let π be a unitary representation of G, and let P be a system of imprimitivity for π based on G/H, H a closed subgroup. Then there exists a representation γ*

of H, *uniquely determined up to equivalence, and a unitary map*

$$U: \mathcal{H}_\pi \to \mathcal{H}(\gamma)$$

such that for all $g \in G$ *and Borel sets* $E \subseteq G/H$ *we have*

$$U\pi(g) = (\mathrm{Ind}_H^G\gamma)(g)U \qquad UP_E = P_E^\gamma U.$$

EXERCISE. In Chapter III, where we use the imprimitivity theorem in a crucial way, the group actions are written on the left. Using a left action formulation of induced representations (see Appendix A, Exercise 5), rework the imprimitivity theorem so as to accommodate left actions.

NOTATION AND TERMINOLOGY

I have only one carte blanche convention -- all groups throughout the text are assumed to be locally compact Hausdorff and *separable*. It saves me trouble.

1. The following symbols denote the sets as indicated:

$\mathbb{R}$ = the real numbers

$\mathbb{C}$ = the complex numbers

$\mathbb{T}$ = $\{z \in C: |z| = 1\}$

$\mathbb{Z}$ = the integers

$\mathbb{Z}_n$ = the integers mod n, $n \geqq 1$

$\mathbb{Q}$ = the rational numbers

$\mathbb{H}$ = the quaternions

$\mathbb{Q}_p$ = the p-adic numbers

k^* = the multiplicative group of non-zero elements in a field k

$\mathbb{R}_+^*$ = the positive real numbers

$\mathbb{Z}_+$ = the positive integers

2. If V is a vector space over a field K, we write

End(V) = the space of K-endomorphisms of V

GL(V) = the non-singular endomorphisms in End(V).

If $\{X_j\}_{1 \leqq j \leqq n}$ is a basis of V, $\dim_K V = n$, then $\{X_j^*\}_{1 \leqq j \leqq n}$ denotes the dual basis of $V^* = \mathrm{Hom}_K(V,K)$.

For $n \geqq 1$

M(n,K) = the n×n matrices with entries in K

GL(n,K) = the non-singular matrices in M(n,K)

D(n,K) = the diagonal matrices in GL(n,K).

If $X \in M(n,K)$, the symbol ${}^tX$ denotes the transposed matrix. The letter I usually denotes the identity operator on whatever vector space is under consideration. $\mathcal{H}$ usually denotes a Hilbert space, $\mathcal{U}(\mathcal{H})$ = the unitary operators thereon.

3. If X is a Borel space, we write

$\mathcal{B}(X)$ = the Borel sets of X.

For locally compact Hausdorff topological spaces X,

$C_0(X)$ = the continuous functions of compact support on X

$C_0^+(X) = \{f \in C_0(X): f(x) \geqq 0 \ \forall x \in X\}$.

If X is a C^∞-manifold

$C_0^\infty(X)$ = the infinitely differentiable functions of compact support.

If G is a locally compact group, e always denotes the identity. Also G^0 = the connected component of G containing e, also called the neutral component,

Δ_G = the modular function of G

Z_G = Cent G = the center of G.

If G is a Lie or algebraic group

$\mathfrak{g}$ = LA(G) is the Lie algebra of G.

We write ad, Ad, Ad* respectively for the adjoint representation of $\mathfrak{g}$, adjoint representation of G on $\mathfrak{g}$, and the co-adjoint representation of G on $\mathfrak{g}^*$.

If X is a Borel G-space

$G_x = \{g \in G: g\cdot x = x\}$ = the stabilizer of x

$G\cdot x = \{g\cdot x: g \in G\}$ = the G-orbit containing x.

4. The phrase algebraic group in the text means linear algebraic group. Zariski (as in Zariski-closed) is usually abbreviated by Z-. For subsets S of Lie or algebraic groups or algebras,

$Z(S)$ = the centralizer of S

$N(S)$ = the normalizer of S.

5. The phrase representation when applied to groups means continuous unitary representation (except in Chapter V, section B). If π is

such a representation, $\mathcal{H}_\pi$ denotes the space on which it acts. Also

$\overline{\pi}$ denotes the conjugate representation in $\overline{\mathcal{H}}_\pi$

$\pi \cong \pi'$ means unitary equivalence

$\pi \approx \pi'$ means quasi-equivalence

$\mathcal{J}(\pi,\pi')$ = the dimension of the space of intertwining operators for π, π'

$\mathcal{J}(\pi,\pi') = 0$ means π and π' are disjoint

$\mathcal{J}(\pi,\pi) = 1$ means π is irreducible

$\mathcal{R}(\pi,\pi')$ = the W*-algebra generated by the algebra of intertwining operators for π and π'

1_G = the trivial representation of G in a space of dimension 1

λ_G = the (left) regular representation of G

$\pi = \mathrm{Ind}_H^G\, \gamma$ = the representation of G induced from H by γ

Rep(G) = the (concrete) space of unitary representations of G

Irr(G) = the subset of Rep(G) consisting of irreducible representations

$\hat{G} = (\mathrm{Irr}(G)/\cong)$, that is the set of unitary equivalence classes of irreducible unitary representations. We often blur the distinction between a representation $\pi \in \mathrm{Irr}(G)$ and its class in $\hat{G}$.

The space $\hat{G}$ carries a natural Borel structure (the Mackey Borel structure) and a natural topology (the Fell or dual topology). These are described in Auslander and Moore [1].

6. Finally if $f \in L_1(G)$, f^* is defined by $f^*(x) = \overline{f(x^{-1})}\Delta(x)^{-1}$, $x \in G$. Also if $\pi \in \mathrm{Rep}(G)$, $f \in L_1(G)$, the symbol $\pi(f)$ denotes the operator

$$\pi(f) = \int_G f(g)\, \pi(g)\, dg.$$

If H is a closed subgroup of G, the symbol $[G:H]$ denotes $\dim L_2(G/H)$.

BIBLIOGRAPHY

At one point the thought occurred to me that since Mackey's survey article in the 1963 Bulletin, no complete bibliography of articles in representation theory has appeared. That's all -- it just occurred to me. I didn't do anything about it. Nevertheless I do hope the reader finds the following list to be of some use. I have included a few references beyond those quoted in the text.

N. Anh

[1] Restriction of the principal series of $SL(n,\mathbb{C})$ to some reductive subgroups, Pacific J. Math., 38(1971), 295-313.

L. Auslander and B. Kostant

[1] Polarization and unitary representations of solvable Lie groups, Invent. Math., 14(1971), 255-354.

L. Auslander and C.C. Moore

[1] Unitary representations of solvable Lie groups, Memoirs Amer. Math. Soc., 62(1966).

V. Bargmann

[1] On a Hilbert space of analytic functions and an associated integral transform I, Comm. Pure and Appl. Math., 14(1961), 187-214.

P. Bernat

[1] Sur les représentations unitaires des groupes de Lie résolubles, Ann. Sci. École Norm. Sup., 82(1965), 37-99.

P. Bernat, N. Conze, M. Duflo, M. Lévy-Nahas, M. Rais, P. Renouard and M. Vergne

[1] Représentations des Groupes de Lie Résolubles, Dunod, Paris, 1972.

R.J. Blattner

[1] On induced representations, Amer. J. Math., 83(1961), 79-98 and 499-512.

[2] Group extension representations, Pacific J. Math., 15(1965), 1101-1115.

A. Borel

[1] Linear algebraic groups, Proc. Symposia Pure Math., 9(1966), 3-19.

[2] Linear Algebraic Groups, Benjamin, New York, 1969.

A. Borel and J.-P. Serre

[1] Théorèmes de finitude en cohomologie galoisienne, Comm. Math. Helv., 39(1964), 111-164.

A. Borel and J. Tits

[1] Groupes réductifs, Publ. I.H.E.S., 27(1965), 55-150.

J. Brezin

[1] Unitary representation theory for solvable Lie groups, Memoirs Amer. Math. Soc., 79(1968).

F. Bruhat

[1] Sur les représentations induites des groupes de Lie, Bull. Soc. Math. France, 84(1956), 97-205.

[2] Sur les représentations des groupes classiques p-adiques I, II, Amer. J. Math., 83(1961), 321-338, 343-368.

C. Chevalley

[1] Sur certaines groupes simples, Tôhoku Math. J., 7(1955), 14-66.

J. Dixmier

[1] Sur les représentations unitaires des groupes de Lie algébriques, Ann. Inst. Fourier, 7(1957), 315-328.

[2] Les Algèbres d'Opérateurs dans l'Espace Hilbertien, Gauthier-Villars, Paris, 1957.

[3] Sur les représentations des groupes de Lie nilpotents I, Amer. J. Math.,81(1959), 160-170.

[4] Sur les représentations des groupes de Lie nilpotents III, Canad. J. Math., 10(1958), 321-348.

[5] Les C*-Algèbres et Leurs Représentations, Gauthier-Villars, Paris, 1964.

[6] Représentations induites holomorphes des groupes résolubles algébriques, Bull. Soc. Math. France, 94(1966), 181-206.

M. Duflo

[1] Sur les extensions des représentations irréductibles des groupes de Lie nilpotents, Ann. Sci. Ecole Norm. Sup., 5(1972), 71-120.

E. Effros

[1] Transformation groups and C*-algebras, Ann. of Math., 81(1965), 38-55.

J.M.G. Fell

[1] The dual spaces of C*-algebras, Trans. Amer. Math. Soc., 94 (1960), 365-403.

[2] A new proof that nilpotent groups are CCR, Proc. Amer. Math. Soc., 13(1962), 93-99.

[3] Weak containment and induced representations of groups, Canad. J. Math., 14(1962), 237-268.

I.M. Gelfand, M.I. Graev and I.I. Pyatetskii-Shapiro

[1] Representation Theory and Automorphic Functions, Saunders, Philadelphia, 1969.

I.M. Gelfand and M.A. Naimark

[1] Unitäre Darstellungen der Klassischen Gruppen, Akadamie-Verlag, Berlin, 1957.

J. Glimm

[1] Locally compact transformation groups, Trans. Amer. Math. Soc., 101(1961), 124-138.

K.I. Gross

[1] The dual of a parabolic subgroup and a degenerate principal series of Sp(n,ℂ), Amer. J. Math., 93(1971), 398-428.

S. Grosser and M. Moskowitz

[1] Harmonic analysis on central topological groups, Trans. Amer. Math. Soc., 156(1971), 419-454.

Harish-Chandra

[1] Representations of semisimple Lie groups III, Trans. Amer. Math. Soc., 76(1954), 234-253.

[2] On a lemma of F. Bruhat, J. Math. Pures et Appl., 35(1956), 203-210.

[3] Representations of semisimple Lie groups VI, Amer. J. Math., 78 (1956), 564-628.

[4] The characters of semisimple Lie groups, Trans. Amer. Math. Soc., 83(1956), 98-163.

[5] Some results on an invariant integral on a semisimple Lie algebra, Ann. of Math. 80(1964), 557-593.

[6] Invariant eigendistribution on a semisimple Lie group, Trans. Amer. Math. Soc., 119(1965), 457-508.

[7] Discrete series for semisimple Lie groups I, Acta Math., 113 (1965), 241-318.

[8] Discrete series for semisimple Lie groups II, Acta Math., 116 (1966), 1-111.

[9] Harmonic analysis on semisimple Lie groups, Bull. Amer. Math. Soc., 76(1970), 529-551.

[10] Harmonic analysis on reductive p-adic groups (notes by G. van Dijk), Lecture Notes in Math., 162(1970).

[11] On the theory of the Eisenstein integral, Lecture Notes in Math., 266(1971), 123-149.

[12] Harmonic analysis on reductive p-adic groups, Conference on Harmonic Analysis on Homogeneous Spaces, Williams College, Williamstown, 1972.

S. Helgason

[1] Differential Geometry and Symmetric Spaces, Academic Press, New York, 1962.

[2] A duality for symmetric spaces with applications to group representations, Advances in Math., 5(1970), 1-154.

T. Hirai

[1] The characters of some induced representations of semisimple Lie groups, J. Math. Kyoto Univ., 8(1968), 313-363.

R. Howe

[1] On the principal series of GL_n over p-adic fields, Trans. Amer. Math. Soc., 177(1973), 275-286.

[2] Kirillov theory for compact p-adic groups, preprint.

[3] Topics in harmonic analysis on solvable algebraic groups, preprint.

J.E. Humphreys

[1] Introduction to Lie Algebras and Representation Theory, Springer Verlag, Berlin, 1972.

H. Jacquet and R.P. Langlands

[1] Automorphic forms on GL(2), Lecture Notes in Math.,114 (1970).

A.A. Kirillov

[1] Unitary representations of nilpotent Lie groups, Russian Math. Surveys, 17(1962), 53-104.

A. Kleppner and R.L. Lipsman

[1] The Plancherel formula for group extensions, Ann. Sci. École Norm. Sup., 5(1972), 71-120.

[2] The Plancherel formula for group extensions II, Ann. Sci. École Norm. Sup., 6(1973), 103-132.

A.W. Knapp and E.M. Stein

[1] Intertwining operators for semisimple groups, Ann. of Math., 93(1971), 489-578.

B. Kostant

[1] On the existence and irreducibility of certain series of representations, Bull. Amer. Math. Soc., 75(1969), 627-642.

R.A. Kunze and E.M. Stein

[1] Uniformly bounded representations III, Amer. J. Math., 89(1967), 385-442.

S. Lang

[1] Algebraic Numbers, Addison-Wesley, Reading, 1964.

R.L. Lipsman

[1] The dual topology for the principal and discrete series on semisimple groups, Trans. Amer. Math. Soc., 152(1970), 399-417.

[2] An explicit realization of Kostant's complementary series with applications to uniformly bounded representations, preprint.

[3] On the characters and equivalence of continuous series representations, J. Math. Soc. Japan, 23(1971), 452-480.

[4] Representation theory of almost connected groups, Pacific J. Math., 42(1972), 453-467.

[5] The CCR property for algebraic groups, Amer. J. Math., to appear.

[6] Algebraic transformation groups and representation theory, to appear.

G.W. Mackey

[1] Imprimitivity for representations of locally compact groups I, Proc. Nat. Acad. Sci., 35(1949), 537-545.

[2] Induced representations of groups, Amer. J. Math., 73(1951), 576-592.

[3] Induced representations of locally compact groups I, Ann. of Math., 55(1952), 101-139.

[4] Induced representations of locally compact groups II, Ann. of Math. 58(1953), 193-221.

[5] Borel structures in groups and their duals, Trans. Amer. Math. Soc., 85(1957), 134-165.

[6] Unitary representations of group extensions I, Acta Math. 99 (1958), 265-311.

[7] Induced representations and normal subgroups, Proc. Int. Symp. Linear Spaces, 319-326, Permagon, Oxford, 1960.

R.P. Martin

[1] On the decomposition of tensor products of principal series representations for real-rank one semisimple groups, Thesis, Univ. of Maryland, 1973.

C.C. Moore

[1] Compactifications of symmetric spaces, Amer. J. Math., 86 (1964), 201-218.

[2] Decomposition of unitary representations defined by discrete subgroups of nilpotent groups, Ann. of Math., 82(1965), 146-182.

[3] Groups with finite-dimensional irreducible representations, Trans. Amer. Math. Soc., 166(1972), 401-410.

[4] Representations of solvable and nilpotent groups and harmonic analysis on nil and submanifolds, Conference on Harmonic Analysis on Homogeneous Spaces, Williams College, Williamstown, 1972.

G.D. Mostow

[1] Fully reducible subgroups of algebraic groups, Amer. J. Math., 78(1956), 200-221.

K. Okamoto

[1] On induced representations, Osaka J. Math., 4(1967), 85-94.

L. Pukanszky

[1] On the Kronecker products of irreducible representations of the 2×2 real unimodular group I, Trans. Amer. Math. Soc., 100(1961), 116-152.

[2] On the theory of exponential groups, Trans. Amer. Math. Soc., 126(1967), 487-507.

[3] Leçons sur les Représentations des Groupes, Dunod, Paris, 1967.

[4] On the characters and the Plancherel formula of nilpotent groups, J. Functional Analysis, 1(1967), 255-280.

[5] On the unitary representations of exponential groups, J. Functional Analysis, 2(1968), 73-113.

[6] Characters of algebraic solvable groups, J. Functional Analysis, 3(1969), 435-491.

[7] Representations of solvable Lie groups, Ann. Sci. École Norm. Sup., 4(1971), 464-608.

S. Quint

[1] Representations of solvable Lie groups, Berkeley Lecture Notes, 1972.

[2] Representation theory of solvable Lie groups, Thesis, Univ. of California at Berkeley, 1973.

R. Richardson

[1] Principal orbit types for algebraic transformation spaces in characteristic zero, Invent. Math., 116(1972), 6-14.

H. Rossi and M. Vergne

[1] Representations of certain solvable Lie groups on Hilbert spaces of holomorphic functions and applications to the holomorphic discrete series of a semisimple Lie group, J. Functional Analysis, 13(1973), 324-389.

P.J. Sally

[1] Analytic continuation of the irreducible unitary representations of the universal covering group of $SL(2,\mathbb{R})$, Memoirs Amer. Math. Soc., 69(1967).

P.J. Sally and J. Shalika

[1] Characters of the discrete series of representations of SL(2) over a local field, Proc. Nat. Acad. Sci., 61(1968), 1231-1237.

[2] The Plancherel formula for SL(2) over a local field, Proc. Nat. Acad. Sci., 63(1969), 661-667.

W. Schmid

[1] On a conjecture of Langlands, Ann. of Math., 93(1971), 1-42.

D. Shale

[1] Linear symmetries of free boson fields, Trans. Amer. Math. Soc., 103(1962), 149-167.

R. Streater

[1] The representations of the oscillator group, Comm. Math. Physics, 4(1967), 217-236.

O. Takenouchi

[1] Sur la facteur représentation des groupes de Lie de type E, Math. J. Okayama Univ., 7(1957), 151-161.

P.C. Trombi and V.S. Varadarajan

[1] Spherical transforms on semisimple Lie groups, Ann. of Math., 94(1971), 246-303.

G. Van Dijk

[1] Computation of certain induced characters of p-adic groups, preprint.

M. Vergne

[1] Étude de certaines représentations induites d'un groupe de Lie résoluble exponentiel, Ann. Sci. Ecole Norm. Sup., 3(1970), 353-384.

N. Wallach

[1] Cyclic vectors and irreducibility for principal series representations, Trans. Amer. Math. Soc., 158(1971), 107-113.

G. Warner

[1] Harmonic Analysis of Semi-Simple Groups, 2 volumes, Springer-Verlag, Berlin, 1972.

F. Williams

[1] Reduction of tensor products of principal series representations of complex semi-simple Lie groups, Thesis, Univ. of California at Irvine, 1972.

Vol. 215: P. Antonelli, D. Burghelea and P. J. Kahn, The Concordance-Homotopy Groups of Geometric Automorphism Groups. X, 140 pages. 1971. DM 16,–

Vol. 216: H. Maaß, Siegel's Modular Forms and Dirichlet Series. VII, 328 pages. 1971. DM 20,–

Vol. 217: T. J. Jech, Lectures in Set Theory with Particular Emphasis on the Method of Forcing. V, 137 pages. 1971. DM 16,–

Vol. 218: C. P. Schnorr, Zufälligkeit und Wahrscheinlichkeit. IV, 212 Seiten. 1971. DM 20,–

Vol. 219: N. L. Alling and N. Greenleaf, Foundations of the Theory of Klein Surfaces. IX, 117 pages. 1971. DM 16,–

Vol. 220: W. A. Coppel, Disconjugacy. V, 148 pages. 1971. DM 16,–

Vol. 221: P. Gabriel und F. Ulmer, Lokal präsentierbare Kategorien. V, 200 Seiten. 1971. DM 18,–

Vol. 222: C. Meghea, Compactification des Espaces Harmoniques. III, 108 pages. 1971. DM 16,–

Vol. 223: U. Felgner, Models of ZF-Set Theory. VI, 173 pages. 1971. DM 16,–

Vol. 224: Revètements Etales et Groupe Fondamental. (SGA 1). Dirigé par A. Grothendieck XXII, 447 pages. 1971. DM 30,–

Vol. 225: Théorie des Intersections et Théorème de Riemann-Roch. (SGA 6). Dirigé par P. Berthelot, A. Grothendieck et L. Illusie. XII, 700 pages. 1971. DM 40,–

Vol. 226: Seminar on Potential Theory, II. Edited by H. Bauer. IV, 170 pages. 1971. DM 18,–

Vol. 227: H. L. Montgomery, Topics in Multiplicative Number Theory. IX, 178 pages. 1971. DM 18,–

Vol. 228: Conference on Applications of Numerical Analysis. Edited by J. Ll. Morris. X, 358 pages. 1971. DM 26,–

Vol. 229: J. Väisälä, Lectures on n-Dimensional Quasiconformal Mappings. XIV, 144 pages. 1971. DM 16,–

Vol. 230: L. Waelbroeck, Topological Vector Spaces and Algebras. VII, 158 pages. 1971. DM 16,–

Vol. 231: H. Reiter, L^1-Algebras and Segal Algebras. XI, 113 pages. 1971. DM 16,–

Vol. 232: T. H. Ganelius, Tauberian Remainder Theorems. VI, 75 pages. 1971. DM 16,–

Vol. 233: C. P. Tsokos and W. J. Padgett. Random Integral Equations with Applications to stochastic Systems. VII, 174 pages. 1971. DM 18,–

Vol. 234: A. Andreotti and W. Stoll. Analytic and Algebraic Dependence of Meromorphic Functions. III, 390 pages. 1971. DM 26,–

Vol. 235: Global Differentiable Dynamics. Edited by O. Hájek, A. J. Lohwater, and R. McCann. X, 140 pages. 1971. DM 16,–

Vol. 236: M. Barr, P. A. Grillet, and D. H. van Osdol. Exact Categories and Categories of Sheaves. VII, 239 pages. 1971. DM 20,–

Vol. 237: B. Stenström, Rings and Modules of Quotients. VII, 136 pages. 1971. DM 16,–

Vol. 238: Der kanonische Modul eines Cohen-Macaulay-Rings. Herausgegeben von Jürgen Herzog und Ernst Kunz. VI, 103 Seiten. 1971. DM 16,–

Vol. 239: L. Illusie, Complexe Cotangent et Déformations I. XV, 355 pages. 1971. DM 26,–

Vol. 240: A. Kerber, Representations of Permutation Groups I. VII, 192 pages. 1971. DM 18,–

Vol. 241: S. Kaneyuki, Homogeneous Bounded Domains and Siegel Domains. V, 89 pages. 1971. DM 16,–

Vol. 242: R. R. Coifman et G. Weiss, Analyse Harmonique Non-Commutative sur Certains Espaces. V, 160 pages. 1971. DM 16,–

Vol. 243: Japan-United States Seminar on Ordinary Differential and Functional Equations. Edited by M. Urabe. VIII, 332 pages. 1971. DM 26,–

Vol. 244: Séminaire Bourbaki – vol. 1970/71. Exposés 382–399. IV, 356 pages. 1971. DM 26,–

Vol. 245: D. E. Cohen, Groups of Cohomological Dimension One. V, 99 pages. 1972. DM 16,–

Vol. 246: Lectures on Rings and Modules. Tulane University Ring and Operator Theory Year, 1970–1971. Volume I. X, 661 pages. 1972. DM 40,–

Vol. 247: Lectures on Operator Algebras. Tulane University Ring and Operator Theory Year, 1970–1971. Volume II. XI, 786 pages. 1972. DM 40,–

Vol. 248: Lectures on the Applications of Sheaves to Ring Theory. Tulane University Ring and Operator Theory Year, 1970–1971. Volume III. VIII, 315 pages. 1971. DM 26,–

Vol. 249: Symposium on Algebraic Topology. Edited by P. J. Hilton. VII, 111 pages. 1971. DM 16,–

Vol. 250: B. Jónsson, Topics in Universal Algebra. VI, 220 pages. 1972. DM 20,–

Vol. 251: The Theory of Arithmetic Functions. Edited by A. A. Gioia and D. L. Goldsmith VI, 287 pages. 1972. DM 24,–

Vol. 252: D. A. Stone, Stratified Polyhedra. IX, 193 pages. 1972. DM 18,–

Vol. 253: V. Komkov, Optimal Control Theory for the Damping of Vibrations of Simple Elastic Systems. V, 240 pages. 1972. DM 20,–

Vol. 254: C. U. Jensen, Les Foncteurs Dérivés de $\varprojlim$ et leurs Applications en Théorie des Modules. V, 103 pages. 1972. DM 16,–

Vol. 255: Conference in Mathematical Logic – London '70. Edited by W. Hodges. VIII, 351 pages. 1972. DM 26,–

Vol. 256: C. A. Berenstein and M. A. Dostal, Analytically Uniform Spaces and their Applications to Convolution Equations. VII, 130 pages. 1972. DM 16,–

Vol. 257: R. B. Holmes, A Course on Optimization and Best Approximation. VIII, 233 pages. 1972. DM 20,–

Vol. 258: Séminaire de Probabilités VI. Edited by P. A. Meyer. VI, 253 pages. 1972. DM 22,–

Vol. 259: N. Moulis, Structures de Fredholm sur les Variétés Hilbertiennes. V, 123 pages. 1972. DM 16,–

Vol. 260: R. Godement and H. Jacquet, Zeta Functions of Simple Algebras. IX, 188 pages. 1972. DM 18,–

Vol. 261: A. Guichardet, Symmetric Hilbert Spaces and Related Topics. V, 197 pages. 1972. DM 18,–

Vol. 262: H. G. Zimmer, Computational Problems, Methods, and Results in Algebraic Number Theory. V, 103 pages. 1972. DM 16,–

Vol. 263: T. Parthasarathy, Selection Theorems and their Applications. VII, 101 pages. 1972. DM 16,–

Vol. 264: W. Messing, The Crystals Associated to Barsotti-Tate Groups: With Applications to Abelian Schemes. III, 190 pages. 1972. DM 18,–

Vol. 265: N. Saavedra Rivano, Catégories Tannakiennes. II, 418 pages. 1972. DM 26,–

Vol. 266: Conference on Harmonic Analysis. Edited by D. Gulick and R. L. Lipsman. VI, 323 pages. 1972. DM 24,–

Vol. 267: Numerische Lösung nichtlinearer partieller Differential- und Integro-Differentialgleichungen. Herausgegeben von R. Ansorge und W. Törnig, VI, 339 Seiten. 1972. DM 26,–

Vol. 268: C. G. Simader, On Dirichlet's Boundary Value Problem. IV, 238 pages. 1972. DM 20,–

Vol. 269: Théorie des Topos et Cohomologie Etale des Schémas. (SGA 4). Dirigé par M. Artin, A. Grothendieck et J. L. Verdier. XIX, 525 pages. 1972. DM 50,–

Vol. 270: Théorie des Topos et Cohomologie Etale des Schémas. Tome 2. (SGA 4). Dirigé par M. Artin, A. Grothendieck et J. L. Verdier. V, 418 pages. 1972. DM 50,–

Vol. 271: J. P. May, The Geometry of Iterated Loop Spaces. IX, 175 pages. 1972. DM 18,–

Vol. 272: K. R. Parthasarathy and K. Schmidt, Positive Definite Kernels, Continuous Tensor Products, and Central Limit Theorems of Probability Theory. VI, 107 pages. 1972. DM 16,–

Vol. 273: U. Seip, Kompakt erzeugte Vektorräume und Analysis. IX, 119 Seiten. 1972. DM 16,–

Vol. 274: Toposes, Algebraic Geometry and Logic. Edited by. F. W. Lawvere. VI, 189 pages. 1972. DM 18,–

Vol. 275: Séminaire Pierre Lelong (Analyse) Année 1970–1971. VI, 181 pages. 1972. DM 18,–

Vol. 276: A. Borel, Représentations de Groupes Localement Compacts. V, 98 pages. 1972. DM 16,–

Vol. 277: Séminaire Banach. Edité par C. Houzel. VII, 229 pages. 1972. DM 20,–

Vol. 278: H. Jacquet, Automorphic Forms on GL(2). Part II. XIII, 142 pages. 1972. DM 16,–

Vol. 279: R. Bott, S. Gitler and I. M. James, Lectures on Algebraic and Differential Topology. V, 174 pages. 1972. DM 18,–

Vol. 280: Conference on the Theory of Ordinary and Partial Differential Equations. Edited by W. N. Everitt and B. D. Sleeman. XV, 367 pages. 1972. DM 26,–

Vol. 281: Coherence in Categories. Edited by S. Mac Lane. VII, 235 pages. 1972. DM 20,–

Vol. 282: W. Klingenberg und P. Flaschel, Riemannsche Hilbertmannigfaltigkeiten. Periodische Geodätische. VII, 211 Seiten. 1972. DM 20,–

Vol. 283: L. Illusie, Complexe Cotangent et Déformations II. VII, 304 pages. 1972. DM 24,–

Vol. 284: P. A. Meyer, Martingales and Stochastic Integrals I. VI, 89 pages. 1972. DM 16,–

Vol. 285: P. de la Harpe, Classical Banach-Lie Algebras and Banach-Lie Groups of Operators in Hilbert Space. III, 160 pages. 1972. DM 16,–

Vol. 286: S. Murakami, On Automorphisms of Siegel Domains. V, 95 pages. 1972. DM 16,–

Vol. 287: Hyperfunctions and Pseudo-Differential Equations. Edited by H. Komatsu. VII, 529 pages. 1973. DM 36,–

Vol. 288: Groupes de Monodromie en Géométrie Algébrique. (SGA 7 I). Dirigé par A. Grothendieck. IX, 523 pages. 1972. DM 50,–

Vol. 289: B. Fuglede, Finely Harmonic Functions. III, 188. 1972. DM 18,–

Vol. 290: D. B. Zagier, Equivariant Pontrjagin Classes and Applications to Orbit Spaces. IX, 130 pages. 1972. DM 16,–

Vol. 291: P. Orlik, Seifert Manifolds. VIII, 155 pages. 1972. DM 16,–

Vol. 292: W. D. Wallis, A. P. Street and J. S. Wallis, Combinatorics: Room Squares, Sum-Free Sets, Hadamard Matrices. V, 508 pages. 1972. DM 50,–

Vol. 293: R. A. DeVore, The Approximation of Continuous Functions by Positive Linear Operators. VIII, 289 pages. 1972. DM 24,–

Vol. 294: Stability of Stochastic Dynamical Systems. Edited by R. F. Curtain. IX, 332 pages. 1972. DM 26,–

Vol. 295: C. Dellacherie, Ensembles Analytiques, Capacités, Mesures de Hausdorff. XII, 123 pages. 1972. DM 16,–

Vol. 296: Probability and Information Theory II. Edited by M. Behara, K. Krickeberg and J. Wolfowitz. V, 223 pages. 1973. DM 20,–

Vol. 297: J. Garnett, Analytic Capacity and Measure. IV, 138 pages. 1972. DM 16,–

Vol. 298: Proceedings of the Second Conference on Compact Transformation Groups. Part 1. XIII, 453 pages. 1972. DM 32,–

Vol. 299: Proceedings of the Second Conference on Compact Transformation Groups. Part 2. XIV, 327 pages. 1972. DM 26,–

Vol. 300: P. Eymard, Moyennes Invariantes et Représentations Unitaires. II. 113 pages. 1972. DM 16,–

Vol. 301: F. Pittnauer, Vorlesungen über asymptotische Reihen. VI, 186 Seiten. 1972. DM 18,–

Vol. 302: M. Demazure, Lectures on p-Divisible Groups. V, 98 pages. 1972. DM 16,–

Vol. 303: Graph Theory and Applications. Edited by Y. Alavi, D. R. Lick and A. T. White. IX, 329 pages. 1972. DM 26,–

Vol. 304: A. K. Bousfield and D. M. Kan, Homotopy Limits, Completions and Localizations. V, 348 pages. 1972. DM 26,–

Vol. 305: Théorie des Topos et Cohomologie Etale des Schémas. Tome 3. (SGA 4). Dirigé par M. Artin, A. Grothendieck et J. L. Verdier. VI, 640 pages. 1973. DM 50,–

Vol. 306: H. Luckhardt, Extensional Gödel Functional Interpretation. VI, 161 pages. 1973. DM 18,–

Vol. 307: J. L. Bretagnolle, S. D. Chatterji et P.-A. Meyer, Ecole d'été de Probabilités: Processus Stochastiques. VI, 198 pages. 1973. DM 20,–

Vol. 308: D. Knutson, λ-Rings and the Representation Theory of the Symmetric Group. IV, 203 pages. 1973. DM 20,–

Vol. 309: D. H. Sattinger, Topics in Stability and Bifurcation Theory. VI, 190 pages. 1973. DM 18,–

Vol. 310: B. Iversen, Generic Local Structure of the Morphisms in Commutative Algebra. IV, 108 pages. 1973. DM 16,–

Vol. 311: Conference on Commutative Algebra. Edited by J. W. Brewer and E. A. Rutter. VII, 251 pages. 1973. DM 22,–

Vol. 312: Symposium on Ordinary Differential Equations. Edited by W. A. Harris, Jr. and Y. Sibuya. VIII, 204 pages. 1973. DM 22,–

Vol. 313: K. Jörgens and J. Weidmann, Spectral Properties of Hamiltonian Operators. III, 140 pages. 1973. DM 16,–

Vol. 314: M. Deuring, Lectures on the Theory of Algebraic Functions of One Variable. VI, 151 pages. 1973. DM 16,–

Vol. 315: K. Bichteler, Integration Theory (with Special Attention to Vector Measures). VI, 357 pages. 1973. DM 26,–

Vol. 316: Symposium on Non-Well-Posed Problems and Logarithmic Convexity. Edited by R. J. Knops. V, 176 pages. 1973. DM 18,–

Vol. 317: Séminaire Bourbaki – vol. 1971/72. Exposés 400–417. IV, 361 pages. 1973. DM 26,–

Vol. 318: Recent Advances in Topological Dynamics. Edited by A. Beck, VIII, 285 pages. 1973. DM 24,–

Vol. 319: Conference on Group Theory. Edited by R. W. Gatterdam and K. W. Weston. V, 188 pages. 1973. DM 18,–

Vol. 320: Modular Functions of One Variable I. Edited by W. Kuyk. V, 195 pages. 1973. DM 18,–

Vol. 321: Séminaire de Probabilités VII. Edité par P. A. Meyer. VI, 322 pages. 1973. DM 26,–

Vol. 322: Nonlinear Problems in the Physical Sciences and Biology. Edited by I. Stakgold, D. D. Joseph and D. H. Sattinger. VIII, 357 pages. 1973. DM 26,–

Vol. 323: J. L. Lions, Perturbations Singulières dans les Problèmes aux Limites et en Contrôle Optimal. XII, 645 pages. 1973. DM 42,–

Vol. 324: K. Kreith, Oscillation Theory. VI, 109 pages. 1973. DM 16,–

Vol. 325: Ch.-Ch. Chou, La Transformation de Fourier Complexe et L'Equation de Convolution. IX, 137 pages. 1973. DM 16,–

Vol. 326: A. Robert, Elliptic Curves. VIII, 264 pages. 1973. DM 22,–

Vol. 327: E. Matlis, 1-Dimensional Cohen-Macaulay Rings. XII, 157 pages. 1973. DM 18,–

Vol. 328: J. R. Büchi and D. Siefkes, The Monadic Second Order Theory of All Countable Ordinals. VI, 217 pages. 1973. DM 20,–

Vol. 329: W. Trebels, Multipliers for (C, α)-Bounded Fourier Expansions in Banach Spaces and Approximation Theory. VII, 103 pages. 1973. DM 16,–

Vol. 330: Proceedings of the Second Japan-USSR Symposium on Probability Theory. Edited by G. Maruyama and Yu. V. Prokhorov. VI, 550 pages. 1973. DM 36,–

Vol. 331: Summer School on Topological Vector Spaces. Edited by L. Waelbroeck. VI, 226 pages. 1973. DM 20,–

Vol. 332: Séminaire Pierre Lelong (Analyse) Année 1971-1972. V, 131 pages. 1973. DM 16,–

Vol. 333: Numerische, insbesondere approximationstheoretische Behandlung von Funktionalgleichungen. Herausgegeben von R. Ansorge und W. Törnig. VI, 296 Seiten. 1973. DM 24,–

Vol. 334: F. Schweiger, The Metrical Theory of Jacobi-Perron Algorithm. V, 111 pages. 1973. DM 16,–

Vol. 335: H. Huck, R. Roitzsch, U. Simon, W. Vortisch, R. Walden, B. Wegner und W. Wendland, Beweismethoden der Differentialgeometrie im Großen. IX, 159 Seiten. 1973. DM 18,–

Vol. 336: L'Analyse Harmonique dans le Domaine Complexe. Edité par E. J. Akutowicz. VIII, 169 pages. 1973. DM 18,–

Vol. 337: Cambridge Summer School in Mathematical Logic. Edited by A. R. D. Mathias and H. Rogers. IX, 660 pages. 1973. DM 42,–

Vol. 338: J. Lindenstrauss and L. Tzafriri, Classical Banach Spaces. IX, 243 pages. 1973. DM 22,–

Vol. 339: G. Kempf, F. Knudsen, D. Mumford and B. Saint-Donat, Toroidal Embeddings I. VIII, 209 pages. 1973. DM 20,–

Vol. 340: Groupes de Monodromie en Géométrie Algébrique. (SGA 7 II). Par P. Deligne et N. Katz. X, 438 pages. 1973. DM 40,–

Vol. 341: Algebraic K-Theory I, Higher K-Theories. Edited by H. Bass. XV, 335 pages. 1973. DM 26,–

Vol. 342: Algebraic K-Theory II, "Classical" Algebraic K-Theory, and Connections with Arithmetic. Edited by H. Bass. XV, 527 pages. 1973. DM 36,–